Plumbing Design
and
Installation Reference Guide

The McGraw-Hill Engineering Reference Guide Series

This series makes available to professionals and students a wide variety of engineering information and data available in McGraw-Hill's library of highly acclaimed books and publications. The books in the Series are drawn directly from this vast resource of titles. Each one is either a condensation of a single title or a collection of sections culled from several titles. The Project Editors responsible for the books in the Series are highly respected professionals in the engineering areas covered. Each Editor selected only the most relevant and current information available in the McGraw-Hill library, adding further details and commentary where necessary.

Hicks • PLUMBING DESIGN AND INSTALLATION REFERENCE GUIDE

Covers the fundamentals of plumbing design and installation for a wide variety of industrial buildings and structures. Culled by Tyler G. Hicks from several McGraw-Hill books.

Hicks • POWER PLANT EVALUATION AND DESIGN REFERENCE GUIDE

Provides concise evaluation and design information for power plants serving many different needs—utility, industrial, and commercial. Culled by Tyler G. Hicks from several McGraw-Hill books and magazine articles.

Johnson & Jasik • ANTENNA APPLICATIONS REFERENCE GUIDE

Includes practical information and guidelines to antenna applications in all areas of communication. Comprised of one full section of Johnson and Jasik's Antenna Engineering Handbook, *Second Edition. Prepared by Richard C. Johnson.*

Markus and Weston • CLASSIC CIRCUITS REFERENCE GUIDE

Collects in one source hundreds of electronic circuits immediately useful in a wide variety of applications. Culled by Charles D. Weston from Markus's Sourcebook of Electronic Circuits, Electronics Circuits Manual, *and* Guidebook of Electronics Circuits.

Merritt • CIVIL ENGINEERING REFERENCE GUIDE

Offers quick reference to major civil engineering fields: structural design, surveying; geotechnical, environmental, and water engineering. A condensation by Max Kurtz of Merritt's Standard Handbook for Civil Engineers, *Third Edition.*

Woodson • HUMAN FACTORS REFERENCE GUIDE FOR ELECTRONICS AND COMPUTER PROFESSIONALS

Presents all essential data on human factors (ergonomics) relevant to the electronics and computer fields. Compiled by Wesley E. Woodson from his Human Factors Design Handbook.

Woodson • HUMAN FACTORS REFERENCE GUIDE FOR PROCESS PLANTS

Makes available to engineers and specialists all essential data on human factors (ergonomics) relevant to the process industries. Compiled by Nicholas P. Chopey from Woodson's Human Factors Design Handbook.

Plumbing Design and Installation Reference Guide

Tyler G. Hicks, P.E., Editor
Principal, International Engineering Associates

McGraw-Hill Book Company

New York St. Louis San Francisco Auckland Bogotá Hamburg
Johannesburg London Madrid Mexico Milan Montreal New Delhi
Panama Paris São Paulo Singapore Sydney Tokyo Toronto

Library of Congress Cataloging-in-Publication Data

Hicks, Tyler, Gregory, date.
Plumbing design and installation reference guide.

(The McGraw-Hill engineering reference guide series)
(McGraw-Hill handbook series)
Includes index.
1. Plumbing. 2. Industrial buildings—Sanitation.
I. Title. II. Series.
TH6515.F2H53 1986 696'.1 86-7378
ISBN 0-07-028788-0

1 2 3 4 5 6 7 8 9 0 DOC/DOC 893109876

ISBN 0-07-028788-0

Printed and bound by R.R. Donnelley & Sons Company.

Contents

Preface

This is a practical design manual aimed at helping anyone designing plumbing systems. Typical users who will find the book helpful include plumbing designers, drafters, consulting engineers, architects, plumbing contractors, special-effects designers (those charged with designing waterfalls, spray ponds for decorative purposes, and similar installations), mechanical engineers, civil engineers, building contractors, construction supervisors, and others in the design and construction fields. Both beginning and experienced designers will find the book useful because it covers a wide range of plumbing design procedures.

The book starts with general design procedures for individuals doing plumbing design for any type of installation. A step-by-step procedure is presented which will save time and energy for anyone planning plumbing work and drawings. Each step tells the designer what to do at a particular stage of the plumbing design. Other topics covered in this section include site work, fire protection, energy conservation in plumbing, and typical water loads to be handled in various types of structures.

To speed calculations and organize both the input and results, Section 2 presents a number of plumbing calculation forms. Developed by an experienced plumbing designer, these forms will help every designer do a better job in a shorter time. Further, the results of the design calculations will be readily available for easy reference and checking, if desired.

Nearly every plumbing system requires that water be supplied to a structure for use by humans. Section 3 shows the designer how to size the water-supply system so its users have sufficient water at the desired pressure when needed. A number of typical designs are shown, using worked-out procedures to assist the designer. Using the data provided, any designer will be able to develop and plan a water-supply system that's efficient and economical.

Hot-water supply systems are analyzed in Section 4. Next to cold-water systems, hot-water design is second in importance. And with energy costing more than ever before, conservation approaches are extremely important to every designer. To meet these needs, the section covers the many factors in the safe and economic design of hot-water systems for a variety of installations.

Once water enters a building, it must be drained away after serving the uses for which it was brought into the building. Section 5 covers drainage system design. Both internal and external drainage systems are discussed in this section. Pipe type and size are considered, to enable the designer to make the correct choice for the system being designed.

Storm drains, to remove the water caused by rainstorms and other natural occurrences, are discussed in Section 6. This section covers the design of storm drains so they are of adequate capacity and properly sized and sloped. Combined drains and sewers are also evaluated, as is continuous flow.

Drainage systems must be sized so they can handle the anticipated flow without overloading. At the same time, the money spent on the system must not be excessive, in terms of an optimum design. This section shows every designer how to size a drainage system so it serves the load both efficiently and economically.

Vent piping prevents an unwanted vacuum from being developed in a plumbing system. Likewise, it carries off gases developed in the system. Section 8 covers the design of vent piping so a plumbing system can operate efficiently and effectively.

Newer structures and facilities often require different plumbing systems from conventional buildings and plants. Section 9 shows the designer how to design many unusual plumbing systems—such as waterfalls, decorative spray ponds, demineralized water systems, swimming-pool systems, compressed-air and vacuum systems, and others. This section alone is worth the price of the book since it provides the designer with tools needed for the unusual and unexpected design assignment.

Interior fire protection is essential for the safety of building occupants and contents. Section 10 provides the basic information a plumbing designer needs to design safe and efficient fire-protection systems for many different types of structures.

Where sewage and drainage must be lifted to a sewer higher than the lowest drain point in a building or other structure, a pump is generally used. Section 11 discusses the pumps used, their characteristics, relative cost, and operating features. With this information on hand, the designer can easily select the best pump for a given application.

Thus, this reference guide provides any designer with valuable and useful information needed to design a safe and efficient plumbing system. The information presented was drawn from a number of outstanding works on plumbing written by highly qualified specialists. The books used as sources are acknowledged in each section of the present work.

The editor is grateful to the original authors for providing so much valuable information from which to draw the content of this book. Where the present material refers to earlier data in the original book, such data has been included at the end of the particular section in this book—if the editor considered it pertinent to the reader. Should readers have any questions they believe the editor can answer, they can contact him in care of the publisher.

Tyler G. Hicks

Plumbing Design
and
Installation Reference Guide

Section 1

General Design Procedures

For efficient design of any plumbing system the designer must have an organized approach to the task. This section provides such an approach, one that has proven successful for hundreds of plumbing designs in a variety of structures, large and small.

The section starts with a step-by-step procedure for the design of plumbing work and the preparation of plumbing drawings. Each step tells the designer what to do at a particular stage of the design. Items to be checked when completing a design project are also listed for the designer.

Next the section covers plumbing, fire-protection, and site systems and how they should be handled by the designer. The typical water loads expected in various types of structures are given, as are the fire-protection requirements. Energy conservation possibilities in plumbing systems are also considered from the designer's view.

Site work, with its important sewer, water, and gas data, is discussed next. From here the section moves to pipe material selection. Then water-supply systems are considered, showing the designer how to work with local authorities to see that the design is properly served.

Using the data in this section the plumbing designer will be well equipped to produce economic and safe designs which meet the requirements of all local codes and the occupants of the building. Both time and money will be saved.

A. Information to be obtained and items to be checked at the start of a project.

1. Location of the building or buildings on the site.
2. Determine size, location, and depth of all adjacent available sewers, water, and gas mains. Investigate the requirements of local departments and/or utility companies, even the minor details, and confirm all information obtained. Failure to go into this thoroughly or to confirm information received can lead to embarrassing controversies when the project is under construction. If the locality is not too far distant, a telephone call to get or recheck information is worthwhile and should be made when necessary.
3. Ascertain available water pressure at a given elevation.
4. If the project is an alteration or an addition to an existing building, check existing services and equipment capacities.
5. Determine the need for a fire standpipe system in the building.
6. Check on the need for sprinklers in the building.
7. Find out the need for special systems such as: gas, compressed air, vacuum air, distilled water, demineralized water, reverse osmosis water, medical gases, or other special systems.
8. Check the local codes, if any.
9. Verify with the electrical project engineer the current characteristics for motors, controls, and heaters, and the maximum size of motors that can be started across the line.
10. Resolve with the HVAC project engineer the availability of steam or hot water for heating the domestic hot water.
11. Obtain a detailed kitchen layout and equipment list.

B. Procedure for the design of plumbing work and the preparation of plumbing drawings.

1. Count the plumbing fixtures and estimate the following:

    a. Sizes of house sewers.
    b. Size of water services.
    c. Size of gas service.
    d. Hot water load (tank size and necessary makeup).

2. Determine if sump pumps or ejector pumps are required.
3. Determine if constant-pressure booster pumps or gravity or pneumatic tanks are required.
4. Note the presence or absence of sufficient slop sinks, fixtures for the handicapped, and drinking fountains or water coolers; and check any lack with the architect.
5. Ascertain whether drinking fountains will be supplied from a central or subcentral chilled water system or whether drinking fountains must be provided with local or integral cooling units.
6. Notify the electrical project engineer of the electrical requirements of the plumbing work.
7. Notify the HVAC project engineer of the domestic hot water heating requirements.
8. Ascertain from the HVAC project engineer his drain, water, and gas requirements. Great care should be exercised to work closely with the project engineer on all projects, so that:

    a. All water and gas requirements are satisfied by valved connections in the plumbing contract within 10 ft developed length of their equipment inlets, as the HVAC contract normally covers only the last 10 ft of piping to the equipment.
    b. Necessary floor drains, funnel drains, etc., are provided reasonably adjacent to all HVAC equipment requiring drains. All drainage piping between the equipment outlets and the floor drains, funnel drains, etc., will normally be provided in the HVAC contract.

9. Locate roof drains, leaders, main stacks, fire standpipe risers, sprinkler risers, etc.
10. Check with the HVAC and electrical project engineers regarding space conditions for risers, fire-hose cabinets, etc., and for required machinery room space.
11. Check the structural drawings for space conditions affecting the plumbing work.
12. Obtain minimum ceiling heights from the architectural drawings or architect and check to see that the plumbing work clears these heights.
13. Note lack of proper pipe spaces and wall thicknesses for the plumbing work and check this lack with the architect.
14. In general, piping should be run as direct as possible. However, except for piping buried underground, all piping should be run parallel to and at right angles to the walls, partitions, etc., and should be neatly grouped in parallel lines.
15. In general piping should be run to clear steel and concrete beams. Where absolutely necessary, piping may be run through beams. Where it is necessary to clip beam flanges or run piping through the web of steel beams or through concrete beams, permission from the structural engineer must be obtained and confirmed; and all such special conditions should be clearly noted on the drawings.
16. Note piping rising within a story as "rise." Note that rising to the story above as "UP." Piping dropping within a story should be noted as "drop." That dropping to the story below should be noted as "Dn." Piping at the ceiling should be noted as "at ceiling" when exposed and as "in ceiling" when concealed. Piping under the floor, other than obvious fixture drain lines, should be noted as "under floor," "at ceiling below," or "in ceiling below," as required.
17. Riser diagrams should be drawn at 1/4-in scale whenever possible (practical), but never less than 1/8-in scale. Water service, hot water heaters, sump pumps, ejector pumps, water pumps, tanks, etc., should be shown on riser diagrams.
18. All three phase motor controllers (starters), remote control stations, alarm panels, remote

alarms, remote pressure switches, float switches, and electrode control units should be located and noted on the drawings.

19. Unless the architect or owner has a standard list of symbols that they request be used, plumbing drawings should use standard office symbols and abbreviations.
20. All information received from or given to the architect, owner, local authorities, etc., either in person or on the telephone, should be immediately confirmed by memorandum with copies sent to individuals involved.
21. When attending project meetings, detailed notes should be taken of all items of discussion pertaining either directly or indirectly to the plumbing work. After the meeting, a memorandum should be written confirming the details discussed with a copy sent to the others present at the meeting.
22. After a project has been scheduled and/or started, the project engineer should record by memorandum the origin of any change or addition to the work, the time involved (man days), and any resultant delay in completion, and have this memorandum approved before proceeding with the work involved.
23. Decisions affecting a trade other than one's own should be avoided when requested by the architect and/or owner. Your answer should be that the information will be obtained from the particular project engineer involved when you return to the office, or if necessary, by calling the project engineer involved from the architect's or owner's office. The same procedure should apply to telephone discussions.

C. Items to be checked when completing a project.

1. North point on all plans.
2. Column numbers, room names, and numbers on all plans.
3. Floor elevations indicated at least on first floor and basement or cellar floor plans.
4. General notes and reference to them on each drawing.

5. Symbol list.
6. Completeness of titles and title block, and presence of PE stamp on all drawings.
7. Reference notes between plans and risers, and between plans and plot (site) plan.
8. Indicated elevations of house sewers leaving the building.
9. Indications of sillcocks around the building on appropriate floor plan and plot plan.
10. Plans against risers for agreement of pipe sizing and location of vent and water branches (high or low).
11. Plumbing plans against those of the heating, ventilating, air-conditioning, and electrical trades for space clearances.
12. That the electrical project engineer has connected up all heaters, motors, controls, alarms, etc., and provided all required heat tracing (to prevent freezing). Check for agreement on location of motor controllers (starters), and horsepower or kilowatts required.
13. That the HVAC project engineer has connected up all hot water heaters, etc.
14. That all HVAC required water connections have been provided and are in the proper locations.
15. That all HVAC required drains have been provided and are in the proper locations.

PLUMBING, FIRE PROTECTION, AND SITE SYSTEMS

A. General.

1. Conformity with requirements.

    a. All work shall be designed in accordance with the requirements of all applicable codes, the requirements of all local authorities having jurisdiction, the requirements of the owner's fire insurance underwriters, and in accordance with latest good engineering practice.
    b. For projects without plumbing code jurisdiction, contact your supervisor for the code to be followed.
    c. The following authorities should be checked

for jurisdiction, availability of services, and requirements:

(1) Building department.
(2) Sewer department.
(3) Water department or water company.
(4) Gas department or gas company.
(5) Fire department.
(6) Public health and/or public safety departments.
(7) Hospital department.

d. All work shall be designed in accordance with this design manual, unless modified by your supervisor.

2. Principles of design.

a. Your supervisor should always be consulted on the design of all unusual items, and any other items or systems which are not an everyday part of plumbing systems.
b. See listing of preliminary plumbing utility loads, item 3.
c. It is imperative to maintain consistency in design and detail from project to project on the same site or for the same owner, because inconsistencies without definite justification are impossible to explain to an owner.
d. When handling buildings in new localities, investigate local customs and local requirements. Not only read the code but try to find out something about local practices. ASSE could be of help in this respect. Many plumbing inspectors are ASSE members.
e. The group designing the site work, i.e., site drainage, treatment plants, water supply plants, and other site utilities, and the sprinkler group should be given as much advance notice as possible regarding the starting and finishing dates and particularly changes in dates on the projects on which they are participating. It should be remembered that they are working simultaneously on several projects

and must coordinate their program in order to accomplish maximum service.

f. For projects on which the owner's standard specifications are used (with or without modifications), make sure that all items normally covered in the specifications are covered for that particular project. If not, they must be added to the specifications or drawings so that there is a complete whole.

g. When an architect insists on changing the wording of the introductory paragraph under "Scope of Work," or any of the "Definitions," the matter should be checked with your supervisor, because failure to do so can lead to serious legal consequences with respect to completeness of the specifications.

h. In writing the list of acceptable manufacturers, care should be taken to select manufacturers not only for the quality of the project involved but also for the acceptability of their product with respect to the specifications. All divergencies from the list of acceptable manufacturers either in initial writing of the specifications or later in submission must be checked with your supervisor. (It is not necessary to list manufacturers for hangers and pipe and fittings where the pipe and fittings are specified by ASTM or ANSI Standards.)

i. Before commencing work on preliminaries, the project should be carefully reviewed with your supervisor to set up the design criteria and the work that should be indicated on the preliminary drawings, because such things vary widely with the type of building and the purpose of the preliminary.

   (1) General tendency has been to show more than is necessary on the preliminaries, which needlessly runs up the job cost.

   (2) The extent of the work and the detail indicated should also be coordinated with the other trades so that all are compatible.

j. The calculations shall be complete, including indications of all assumptions made and notations of the origin of outside figures appearing in them. In other words, they should be so complete and detailed that they stand on their own merits and can be interpreted by anyone without an explanation from the writer.

   (1) All calculations shall be dated.
   (2) Void calculations should be clearly marked "void" when a new calculation is made.

k. Special form calculation sheets should be used. See Section 2.

   (1) One-half to two-thirds of the data is printed on the calculation sheets, which will save time.
   (2) The calculation sheets proceed in a step-by-step process, thereby eliminating the possibility of leaving out a step because of some distraction.

l. In doing a preliminary, calculations should be prepared on the maximum size of the basic systems, and used for the sizing of the major equipment, even though the basis for such work may be very approximate or hypothetical. This is important for showing the relation between the sizes arrived at in the preliminary and the sizes developed in the final design.

m. On taking over a project that has been started by another, the project engineer should check very carefully through the previous records of the project in order to acquaint himself with what has been done and said about the project previously, to be sure that there is no gap in the work of the two project engineers, i.e., as a result of both assuming that the other took care of a certain item. This is particularly important on government jobs, where specific formal submissions of varying items must be made, and a source of embarrassment if an item is forgotten.

n. Drawing production should be carefully planned in advance to avoid costly overdrafting and to improve the clarity of the drawings. Noting should be kept to a necessary minimum, remembering that riser diagrams supplement floor plans; and if a combination of the two clearly presents the picture, much of the noting on the floor plans is unnecessary. When a pipe runs in the normal manner that such a pipe does in normal plumbing, it is not necessary to note this. However, when a pipe runs abnormally, such a fact must be noted.

o. Structural information should always be transmitted to the structural engineer.

   (1) All verbal information should be confirmed in writing.
   (2) Information given should include sketches for pits, steelcuts, weights of equipment and size of bases, slab openings, etc. (Freehand sketches are good enough.)

p. As automatic 1/2-hp three-phase starters are disproportionately expensive compared with the cost of the equipment controlled, and in many cases it is impossible to get three-phase motors on some of this equipment, the electrical project engineer should allow plumbing systems to use single-phase for 1/2-hp motors.

q. The "Electrical Connection Data Sheet" establishes the general voltages available on the project. See Calculation Form A, Section 2.

   (1) These should be specified for heaters, solenoid valves, relays, etc.
   (2) However, motors should be specified in accordance with the following NEMA standard motor voltages given in the table on the next page.

r. When discussing any major items with the architect or the contractor after a project is in construction, ascertain whether the same

Motor voltage rating, V	System voltage
115	110 to 120
200	208 to 220
230	240
460	480
575	600

questions have been previously discussed with the field observer. This is important so that they can not keep discussing an item with different people in the office until they get a satisfactory answer. Field observers should inform you of all major discussions and interpretations given the architect or contractor. Similarly, project engineers should inform the field observers of any discussions or instructions given the architect or contractor.

s. This design guide does not cover special earthquake design criteria. For such requirements, refer to special design criteria for the earthquake area involved.

3. Preliminary plumbing utility loads. When required at the conceptual stage of a project, preliminary load requirements may be calculated as follows.

   a. Domestic water supply and sanitary drainage. Multiply the population by the following gallons per day (gpd) per person, <u>except</u> as otherwise noted.

Building type	Total water	Hot water
Apartment (average occupancy 2) (per apartment)	150	75
Residence (average occupancy 4) (per residence)	250	125

(Continued on next page)

(Continued)

Building type	Total water	Hot water
Dormitories (residence halls)	65	20
Office building (8-h worker)	20	3
Industrial plant (8-h shift)	35	15
Schools without cafeteria or showers (per student)	15	2
Schools with cafeteria, no showers (per student)	20	4
Schools with cafeteria and showers (per student)	25	10
Hospitals (per bed)	300	125
Cafeteria and restaurants (per meal)	10	4
Places of assembly	3	1/2
Hotels (per room)	75	50
Nursing homes (per bed)	75	30
Airports (per daily passenger)	5	1
Commercial laundry* (per pound of wash)	3.5	2

*Hospital, 15 lb/day/bed; hotel with restaurant, 12 lb/day/bed; hotel without restaurant, 8 lb/day/bed; nursing home, 7 lb/day/bed; apartment or residence, 14 lb/week/living unit.

Temperature of use of hot water: 120°F; except cafeterias and restaurants-140°F, and commercial laundries-180°F.

To obtain the total daily requirements, add cooling tower and other HVAC water requirements to the above.

To obtain the average flow rate (gpm), divide the gallons per day by 60x, the number of hours of normal building operation (use).

To obtain the peak flow rate (gpm), multiply the average flow rate by 2 1/2.

b. Fire flow requirements (gpm).

Buildings	Sprinklers and hose	Hose only	Duration
Light hazard	500	500	30 min
Ordinary hazard:			
Group 1	700	500	60 min
Group 2	850	500	60 min
Group 3	1,250	500	60 min

Light hazard: Office buildings, hospitals, schools, institutional buildings, nursing homes, residential, places of assembly, small libraries, dining rooms.

Ordinary hazard group 1: Parking garages, light-industrial, laundries.

Ordinary hazard group 2: Large libraries, commercial, light industrial.

Ordinary hazard group 3: Exhibition halls, repair garages, warehouse.

Assume a fire in only one building. Assume the building with the greatest requirements.

Add to building requirements 1,000 gpm for outside hose streams.

c. Gas.

(1) For cooking in apartments and residences, use the table at the top of the next page.
(2) For restaurants and laboratories, proportion the requirements from the calculations of a similar project.
(3) Or for restaurants use the second table on the next page.

4. Energy conservation possibilities in plumbing systems.

a. Use of 1/2 gpm lavatory faucet (adaptors for existing buildings).

Living unit or apartment	Percent	Btu/h
1	100	65,000
5	45	146,250
25	25	406,250
50	20	650,000
100	15	975,000
200	13	1,690,000
300	10	1,950,000

To obtain CFH/h for natural gas, divide by 1,000 Btu/ft^3 and for liquified petroleum (LP) gas, divide by 2,500 Btu/ft^3.

Add for hot water heaters 50,000 Btu/h and for gas dryers 35,000 Btu/h.

Type of facility	Btu/customer served
Coffee shops (counter only)	28,500
Coffee shops (table service)	30,168
Cafeteria	32,667
Regular restaurant	33,528

b. Reducing hot water temperature at lavatory faucets from 105° to 95°.
c. Single lavatory faucet with 1/2 gpm flow of tempered water (95°).
d. Use of siphon jet type urinals instead of blowout type (may require privacy shields by the architect between urinals).
e. Use of flush valves on water closets giving 4-gal flush instead of 4 1/2-gal flush.
f. Shutting down hot water circulating pumps at night and over weekends.
g. Bypassing central chilled water unit when incoming water is below 50°.
h. Shutting down chilled water circulating pumps at night and over weekends.

i. Adjusting central chilled water unit to deliver 50° instead of 45° water.
j. Adjusting water coolers to deliver 55° water instead of 50° water.
k. Selecting pumps that have considerable running time, with efficiency the main and cost the secondary consideration.
l. Use of gravity house tank and house pumps instead of constant-pressure booster pumps.
m. Design of display fountains with smaller displays or designing with small displays for everyday use and bigger displays for use on only special occasions.
n. Omit winterizing of display fountain. Shut down and drain in winter.
o. Use of ejector pumps instead of pneumatic ejectors (higher maintenance may result).
p. Use of waste heat preheaters before hot water heaters whenever possible.
q. Use of recirculating water systems on rotary liquid ring type air compressors and vacuum pumps.
r. Use of water-saver tank type water closets (uses 30 percent less water).

SITE WORK

A. General.

1. Principles of design.

a. The site designer should work closely with the building project engineer in establishing the building loads.
b. In sizing the various systems make ample provisions for anticipated future buildings, additions to buildings, and additional paved areas.
c. Connections to public mains and sewers must be made in accordance with requirements of local authorities and codes.
d. Insist that the architect obtain accurate survey information before attempting to finalize outside work, because it is impossible to design this work correctly without accurate

information on existing conditions and final grades.

e. In routing the building utilities, determine if there are any utility interferences between the building and street pipe that you intend to connect to. (In taking over a job from someone else, do not assume that this has already been done; make sure that your routing is clear.)

f. Indicate all existing underground work that is known in the area. This includes piping, valves, manholes, pits, electrical wiring, telephone, etc., whether being connected to or not. This is necessary to avoid problems in the field with regard to the relative location of the pipes being connected to.

g. The requirements of local departments and utility companies should be very carefully investigated and worked out with the respective parties and confirmed in writing.

    (1) It is very important that all their requirements, even minor details, be confirmed with them and not guessed at, in order to avoid unpleasant controversies when the project is under construction.
    (2) Use the utility data sheets on pages 1-18, 1-19, and 1-20.

h. Coordinate the routing of all outside utilities before work is placed on final drawings. This includes sewers, water mains, gas mains, electric and telephone work, steam, high temperature and chilled water, etc.

i. Draw profiles for all storm and sanitary sewers. This improves the layout and also facilitates coordination with other utility crossings.

j. Where existing utilities are located in an area which will be regraded, determine if the utilities will have enough cover, be exposed, or even be under the structure. (Manhole covers, drainage grates, and valve boxes must be adjusted to the new grades. This should be indicated on the drawings.)

Job: ______________________________

Date: ____________________

Sewer Data

1. Size, location, and depth (including manhole locations) of all sewers fronting on or serving the site? Sanitary? Storm water? Combined?

2. If no sewers, how to dispose of sanitary sewage? Storm water drainage? If dry wells or sewage disposal system, topographical map and type of soil at varying depths?

3. If street sewer manholes are required, by whom? If by sewer department, at whose cost? If by owner, any special standard detail?

4. Any special standard detail required for house sewer manholes?

5. Are street sewer connections preferred at or between manholes? If between manholes, are there existing spurs? Are there any minimum or maximum limitations on size or depth?

6. Any other special requirements?

Job: ______________________________

Date: ____________________

Water Data

1. Size, location, and depth of all water mains and fire hydrants fronting on the site or serving the site?

2. Water pressure at a given elevation?

3. Water meter: by water company? Purchased from water company or by owner? Type? Strainer? Water meter pit? If so, by whom? Standard pit details?

4. Fire hydrants: by water company or by owner? Type?

5. Amount of work done by water company? Amount of work done by water company and paid for by the contractor?

6. Minimum cover required on water lines?

7. Any other special requirements?

8. Water analysis.

9. Type of pipe, fittings and joints normally used? Required?

Job: ______________________________

Date: ______________________

Gas Data

1. Size, location and depth of all gas mains fronting on the site or serving the site?
2. Btu content and pressure of gas available?
3. Amount of work done by the gas company? Amount of work done by the gas company and paid for by the contractor?
4. Equipment furnished by the gas company and to be installed by the contractor?
5. Approximate load: Btu/h
6. Size of service?
7. Size of meter?
8. Any special gas company requirements regarding location of meter or others?
9. Type of pipe, fittings, and valves normally used? Required? Coating? Cathodic protection?

B. Drainage systems.

1. General.

a. Principles of design.

(1) At the start of a project, contact the sewer department of the community involved for the availability of storm, sanitary, and combined sewers in the area. (Obtain drawings showing locations, sizes, and depths of all available sewers, also any rules and requirements connected with sewer work.)
(2) All drainage should be by gravity, wherever possible.
(3) Do not design combined site sewers.
(4) Design piping carrying building drainage in accordance with the requirements of applicable codes.
(5) Design storm and sanitary sewers, not covered by code, in accordance with the requirements of all authorities having jurisdiction, or in the absence of same, as hereinafter indicated.
(6) Provide manholes as follows:

    (a) For every change in direction
    (b) For every change in pipe size.
    (c) For every change in pipe pitch (slope).
    (d) For every major connection.
    (e) At least every 300 ft.

(7) Where sewers of different pipe size enter or leave a drainage structure, lower the crown of the larger pipe to match the crown of the smaller pipe. Where this is not possible, check with your supervisor.
(8) Where pipes of equal size enter and leave a manhole, the outlet pipe should be lowered 0.1 ft.
(9) Drop manholes should be provided when

the difference in inlet and outlet elevations is at least the distance required for the required fittings, and is at least 3-1/2 ft.

(10) Piping must be of sufficient strength for the depth of cover. (See rating sheets, Tables 1-1 to 1-3.)

(11) See the following for pipe carrying capacities:

(a) Kutter's pipe-flow chart, Figure 1-1.

(b) Manning's pipe-flow chart, Figure 1-2.

2. Storm water sewer systems.

a. Principles of design.

(1) Obtain building requirements (both initial and ultimate) from the building project engineer. (Make ample provision in the piping for anticipated future building expansion.)

(2) All building roofs, plazas, paved areas, and unpaved areas as required shall be drained by gravity to public sewers or other approved means of disposal. (Where gravity flow is impossible and pumping may be required, check with your supervisor.)

(3) Where storm sewers are not available, the alternate means of disposal to be considered are:

(a) Streams, lakes, or ponds.

(b) Recharge basins.

(c) Dry wells.

(d) On grade.

(4) There are computer programs available which can size the system.

(5) Base storm-water drainage systems on at least a 10-year storm, for privately owned projects. For municipal work, verify this with the proper authorities.

Table 1-1 Pipe Availability for Sewers

Material	Size 1¼	1½	2	2½	3	4	5	6	8	10	12	14	15	16	18	20	21	22	24	27	30	33	36	42	48
Steel	X	X	X	X	X	X	X	X	X	X	X	X		X	X	X			X		X		X	X	X
CI* soil caulked			X		X	X	X	X	X	X	X		X												
DI* water					X	X		X	X	X	X	X		X	X	X			X		X		X	X	X
VCP*						X		X	X	X	X		X		X		X		X	X	X	X	X		
ACP*						X	X	X	X	X	X	X		X	X	X			X	X	X	X	X		
Plain concrete								X	X	X	X		X		X		X		X						
Reinforced concrete											X		X		X		X		X		X	X	X	X	X

*CI: cast iron.
DI: ductile iron.
VCP: vitrified clay pipe.
ACP asbestos cement pipe.

Note to designer:

Steel: used only inside buildings, not below ground.
CI soil: can be used for all conditions. Lengths 5 or 10 ft.
DI water: can be used when CI is required over 15 in in size. Centrifugal cast 3 to 24 in 18 ft long.
VCP: lengths: 3 to 6 ft according to locality.
ACP:

Size, in	Lengths
4	5 ft class 1500; 5 ft and 10 ft class 2400 and 3300
5–6	5 ft and 10 ft class 1500, 2400 and 3300
8	10 ft class 1500; 13 ft class 2400 and 3300
10–36	13 ft all classes

Plain concrete: Lengths up to 8 ft.
Reinforced concrete: Lengths: standard
12 to 24 in; 4 or 6 ft
24 in; 4, 6, or 8 ft

Table 1-2 Site Sewer Pipe Selection Schedule*

Pipe size, in.	Feet of cover: 2 to 2.49	2.5 to 3	3+ to 4	4+ to 5	5+ to 6	6+ to 7	7+ to 8	8+ to 9	9+ to 10	10+ to 12	12+ to 14	14+ to 16	16+ to 18	18+ to 20
6	A	a B	a	a	a	a	a B	a B	A	A	A	A	A	A
8	g B E	E f	b E f	b E f	b E f	b E f	f E	f E A	f B E	f B g E	g B E	g B E	g B F	g B F
10	A g B F	a B f E	a B f E	a B f E	a B f E	a B f E	a B f E	a B f E	A g B E	A g B F	A g B F	A g B F	A g B F	A g B F
12	AC d BF	A c B g E	a B c E f	a bf E c	a B c E f	a B c E f	c A f B E	c C g B F	c A g B F	d AC g BF	d AC g BF	d AC g BF	d AC g BF	d AC h BF
14	h G	g F	g E	g E	g E	g E	g E	g E	g F	g F	h F	h F	h F	h F
15	A c B	A c B	a b c	a b c	a b c	a b c	a B c	a c B	A c B	A c B	A c B	A d B C	A d B C	A d B C
16	h G	g F	g E	g E	g E	g E	g F	g F	g F	h F	h F	h F	j F	j F
18	c A h B F	c A h B F	a bg F c	a bg F c	a bg F c	a bg F c	a bg F c	a bg F c	c A h B F	c A h B F	c A j B F	c A j B F	d AC j BG	d AC j BG
20	h F	h F	g F	g F	g F	g F	g F	g F	h F	h F	j F	j F	j G	j G

	Feet of cover													
Pipe size, in.	2 to 2.49	2.5 to 3	3+ to 4	4+ to 5	5+ to 6	6+ to 7	7+ to 8	8+ to 9	9+ to 10	10+ to 12	12+ to 14	14+ to 16	16+ to 18	18+ to 20
21	A c B	A c B	a b c	a b c	a b c	a b c	a b c	a b c	a b c	A c B	A c B	A c B	A c B	A d B C
24	c A j B G	a B c F h	a bg F c	a bg F c	a bg F c	a bg F c	a bh F c	a bh F c	a B c F h	c A j B F	c A j B G	c A j B G	A c B G	A c B G
27	c A j G	a c G j	a c G h	a c G h	a c G h	a c G h	a c G h	a c G h	a c G j	c A j G	c A j G	A c G	A c G	A c G
30	A c G	a c H j	a c G h	a c G h	a c G h	a c G h	a c G h	a c G j	a c G j	c A j G	A c G	A c G	A c H	A c H
33	A c H	a c H	a c H j	a c H j	a c H j	a c H j	a c H j	a c H j	a c H j	A c H	A c H	A c H	A c H	A c J
36	A c H	a c H	a c H j	a c H j	a c H j	a c H j	a c H j	a c H	a c H	A c H	A c H	A c H	A c J	A c J
42	c	c	c	c	c	c	c	c	c	c	c	c	c	c
48	c	c	c	c	c	c	c	c	c	c	c	c	c	c

*For letter designations see Table 1-3.

Table 1-3 Pipe Material

	Designation	
	Without cradle	With cradle
Vitrified clay pipe - extra strength (VCP)	a	A
Concrete pipe - extra strength (CP)	b	B
Reinforced concrete pipe - class IV (RCP)	c	C
Reinforced concrete pipe - class V (RCP)	d	D
Asbestos cement pipe - class 1500 (ACP)	e	E
Asbestos cement pipe - class 2400 (ACP)	f	F
Asbestos cement pipe - class 3300 (ACP)	g	G
Asbestos cement pipe - class 4000 (ACP)	h	H
Asbestos cement pipe - class 5000 (ACP)	j	J

Notes:

Dead loads: Marston's formula for loads on rigid conduits in trench conditions: $W_c = C_d W B_d^2$; W = 120 /CF; Ku =0.165 max.

Live loads: Holl's integration of Boussineso's formula: Wsc = CsPF/L; Impact factor F = 1.5, H-20 loading; L = 3.01; P = 20,000.

Trench: pipe bedding, class C, shaped bottom, load factor 1.5.

Trench: pipe bedding, class A, concrete cradle, load factor 2.8.

Safety factors: 1.0, reinforced concrete pipe; 1.5, vitrified clay pipe, concrete pipe, asbestos cement pipe.

Maximum allowable load on pipe: ASTM, three-edge bearing method to produce 0.01-in crack.

Trench width: OD, + 2 ft.

All pipe used between structures shall be the strongest pipe required by depth of cover as shown in Table 3-2 unless otherwise indicated.

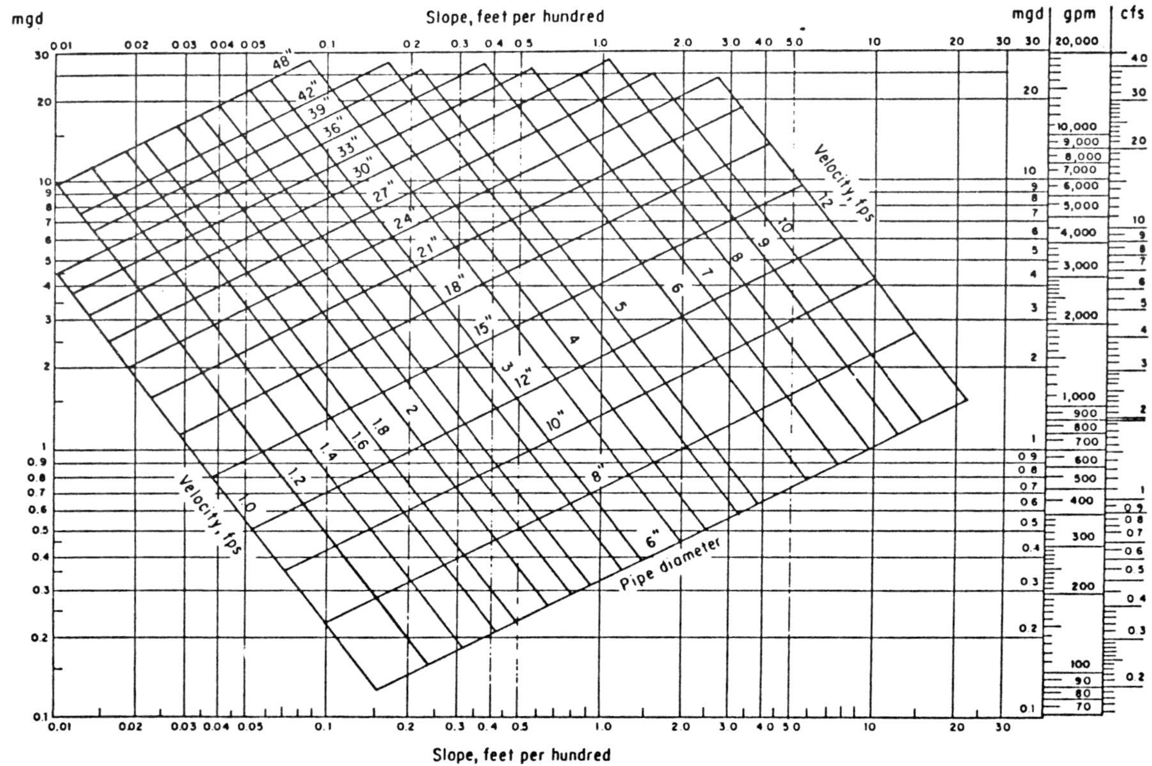

Fig. 1-1. Kutter's pipe flow chart. (N = 0.013.) From Technical Manual TM 5-814-1, Sanitary Engineering, Sanitary and Industrial Waste Sewers, August 1966, by the U.S. Department of the Army.)

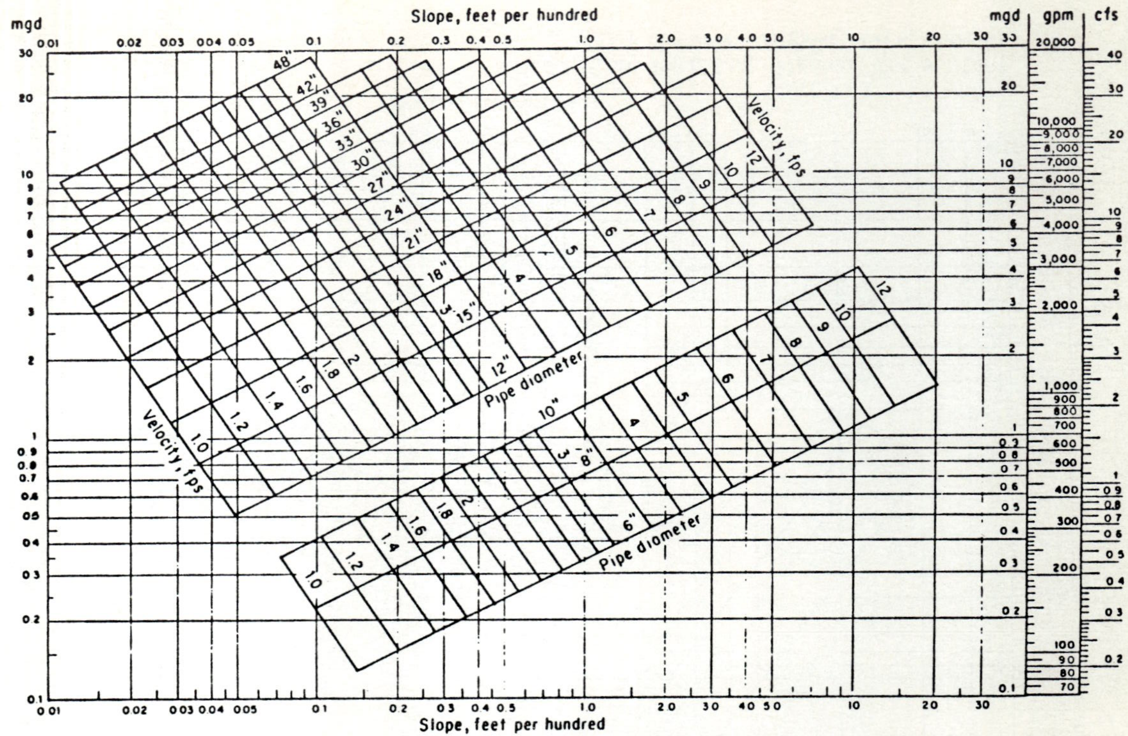

Fig. 1-2. Mannings's pipe flow chart. (N = 0.013 for 12-in pipe and larger; N = 0.014 for 10-in pipe and smaller.) From Technical Manual TM 5-814-1, Sanitary Engineering, Sanitary and Industrial Waste Sewers, August 1966, by the U.S. Department of the Army.)

(6) Piping carrying solely site drainage can be sized using the following formula:

$$Q = ACI$$

where Q = flow, ft^3/s

A = area, acres

I = rainfall rate (use 10-year storm), in/h

C = surface runoff factor

Roofs: $C = 1.00$
Pavement: $C = 0.90$
Grass: $C = 0.30$

(7) See Kutter's pipe-flow chart (Figure 1-1) or Manning's pipe-flow chart (Figure 1-2) for pipe carrying capacity.

(8) Minimum size for site drains is 10 in, with size increased as required by the area drained.

(9) Select pipe size and pitch to provide a minimum velocity of 2 1/2 fps.

(10) Site drainage piping must have minimum cover of 36 in.

(11) Site sewers serving site drains only may be looped through drainage inlets or catch basins (maximum of three) in lieu of manholes. Drainage from buildings should not be looped through drainage inlets or catch basins.

(12) The last drainage structure before connection to a combined sewer must be a hooded type.

(13) All curb-type drains must be hooded catch basins.

(14) Drains in unpaved areas may be catch basins or drainage inlets.

(15) Drains in paved areas must be catch basins.

(16) In working out trench drains with the architect, take care that trenches are wide enough to permit the installation of

a proper type and size drain for its service. If the architect is unwilling to make the trenches the required widths, insist on the use of a cast-iron sectional trench with a drain in the bottom.

(17) Use round grates in grass areas. They are also suitable in paved areas, except where against a curb. They have the advantage of not being dropped into the catch basin or drainage inlet.

(18) Provide all low points on the site with a catch basin or a drainage inlet.

(19) Design piping so that the shortest routing is considered together with the depth of cut.

(20) Avoid placing manholes on very deep piping; try to locate the piping or manholes so that depths do not exceed 15 ft.

(21) Evaluate manhole sizes when multiple lines enter it.

(22) Consider direction of flow when connecting to manholes.

(23) Avoid placing pipes under known locations of trees. Review planting plan during design.

(24) Where possible, avoid placing piping under pavements.

b. Piping.

(1) Piping, unless otherwise noted, should be cast-iron soil pipe and fittings, or class 100 cast-iron or thickness class 2 ductile-iron bell and spigot water pipe and 250-psi fittings.

(2) Piping beyond 8 ft outside the building wall may be extra strength vitrified clay bell and spigot sewer pipe and fittings, extra strength or reinforced-concrete bell and spigot, and/or tongue and groove sewer pipe and fittings, or asbestos cement sewer pipe and fittings, as allowed by code.

(a) Minimum size of nonmetallic piping shall be 8 in.

(3) Use calculation forms C-1, C-2, and C-3, in Section 2, for sizing piping, when computer program is not used.

3. Sanitary sewer systems.

a. Principles of design.

(1) Obtain building requirements (both initial and ultimate) from the building project engineer. (Make ample provision in the piping for anticipated future building expansion.)
(2) All building house sewers should drain by gravity to public sewers or other approved means of disposal.
(3) Where sanitary or combined sewers are not available, alternate means of disposal to be considered are:

(a) Sewage treatment plant.
(b) Septic tank and tile field or leaching cesspools.

(4) Where other than public sewer disposal is required, consult your supervisor on the requirements for this work.
(5) Piping should have a minimum cover of 36 in.
(6) Pipe size and pitch should be selected to provide a minimum velocity of 2 fps.
(7) Sewage pumping stations should be concrete structures (precast or poured in place), normally of the wet-pit-dry-pit type.
(8) Automatic air relief valves should be provided in trapped high points in the pump discharge piping and force mains.

b. Piping.

(1) Piping, unless otherwise noted, should be cast-iron soil pipe and fittings.
(2) Piping beyond 8 ft outside the building wall may be extra strength vitrified clay bell and spigot sewer pipe and fittings,

or asbestos cement sewer pipe and fittings, if allowed by code. (Minimum size of nonmetallic piping shall be 8 in.)

(3) Sewage pump discharge piping inside the station should be galvanized standard-weight steel pipe with threaded cast-iron drainage fittings.

(4) Underground sewage pump discharge piping and force main should be coated cast-iron or ductile-iron bell and spigot water piping, or asbestos cement pressure pipe with cast-iron fittings, if its use is common practice in the area.

c. Equipment.

(1) Sewage ejector pumps.

(a) Pumps should be vertical wet-pit-dry-pit type or submersible type of the Flush-Kleen design.

(b) Pumps should normally be duplex (each full size). In large installations, and to provide for future loads, more than two pumps may sometimes be required.

(c) Calculate the pump head by adding the lift from low water in the pit to 1 ft above grade at the point of discharge (to ensure entry into a surcharged sewer) or the highest elevation of the discharge line, whichever is greater, and the discharge pipe friction (including both individual pump discharge assembly and the common discharge piping).

(d) Low water in the pit is normally 18 in above the bottom for submerged pumps and dry-pit pumps, and the top of the motor for submersible pumps.

(e) Check the friction in the common discharge piping for both one- and two-pump operations. Normally friction for one-pump operation should be included in the specified pump head, provided there is not too great a drop in capacity with the friction of two pumps operating.

(f) The wet pit should have approximately 3 ft gross working depth (this will allow for controls and a reasonable operating cycle).

(g) Sewage ejectors should be sized to handle the peak anticipated flow as calculated from fixture units and National Bureau of Standards Nomograph 31, Curve 1, as shown in Figure 1-3.

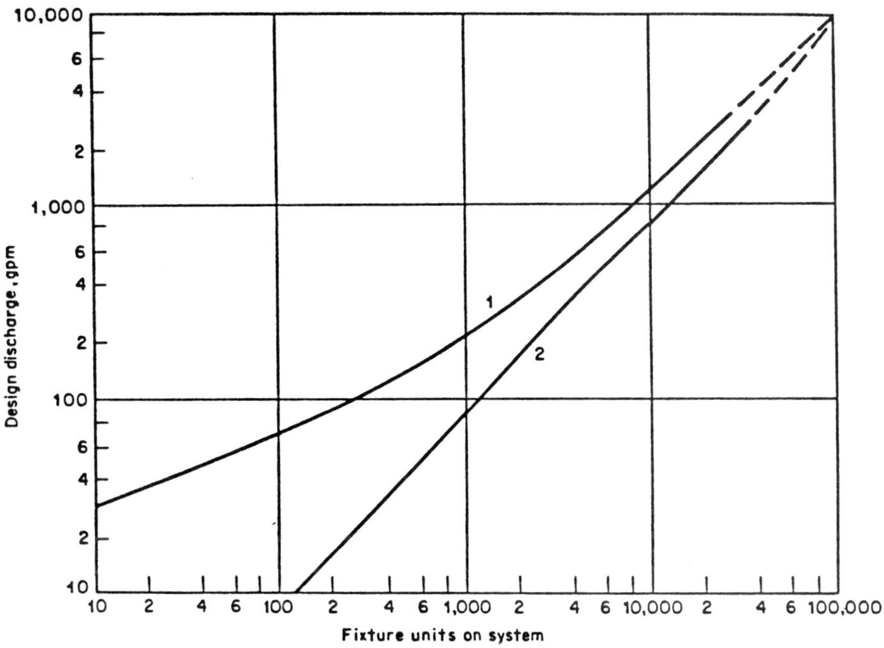

Fig. 1-3. Design-flow curves for plumbing-drainage systems. Curve 1 represents the peak discharge into the drainage system which will not be exceeded more than 1 percent of the time during periods of heaviest use. Curve 2, shown for purposes of comparison only, illustrates the average discharge into the drainage system during periods of heaviest use computed from the discharge characteristics of water closets. (Reprinted by permission from "Air Conditioning, Heating and Ventilating," now "Building Systems Design".)

(h) All controls should be the sealed electrode type, for each pump, or the "bubble" type.
(i) Compression tube high-water alarm actuating units should be provided.
(j) Where the lift is 35 to 40 ft or more, provide ejector pumps with external spring-type check valves instead of the regular swing checks used elsewhere.

C. Water supply systems.

1. General.

a. Principles of design.

(1) The local water department or water company should be contacted for the availability of a public water supply in the area.

(a) Obtain drawings showing locations, sizes, and valving of water mains in the area and available pressure.
(b) Obtain information about the water analysis (if necessary), the type and material of piping and joints, the type of meter, the location of the meter, and what work is normally done by the water department or water company.
(c) In discussion with the local water department or water company, take care to ascertain the maximum possible street pressure as well as the minimum street pressure. The system design should be based on the minimum street pressure; and working pressures of equipment should be based on the maximum possible street pressure.
(d) All projects involving street pressure sprinkler systems, street

pressure fire standpipe systems, fire pumps, or domestic booster pumps require hydrant tests on the mains in streets that could be used to feed the site. Because these tests take time to get, they should be initiated as soon as possible at the start of a project. (Have a hydrant flow test made by the local water department or water company or the underwriters.)

(2) Where public water mains are available, extend service(s) into the site from the mains.

(3) Provide water services (both domestic and fire) to all buildings and to on-site fire hydrants as required.

(4) Sites having hospitals, laboratory-use buildings, or boiler plants should be provided with two services, feeding from separate mains wherever possible. If two separate mains are not available, on-site storage must be provided. Also, large projects such as office buildings, housing, and schools should be similarly provided with two sources of water.

(5) Where no public mains are available, water supply must be developed from wells, streams, or lakes with treatment plants as required and on-site storage of treated water. Consult your supervisor about requirements for this work.

(6) When a public water supply is not available, and wells are being considered, contact a local well driller for recommendations about possible capacities, quality, depths, types, and suggested locations.

(a) There are basically two types of wells: one in unconsolidated (sand) formations having a double casing and a screen; and one in consolidated (rock) formations having a double casing and no screen. See AWWA Standards.

(b) Well pumps are water-lubricated multi-stage turbine (actually multistage diffuser) pumps with electric-motor or engine drive above, or with electric submersible drive.

(c) Well pumps should be provided with electrode low water cutoff.

(d) The pump head is calculated using the lift from the pumping water level, plus on allowance for future drop in level.

(e) The water level in the well is usually much higher when the pump is not operating. If this high water level is below the top intermediate shaft bearing, provide a prelubricating device.

(7) Design potable water supply piping so that no used, unclean, polluted, or contaminated water can enter any portion of such piping from any tank, receptacle, equipment, or plumbing fixture by reason of back siphonage, suction, back pressure, or any other cause, either during normal use and operation thereof, or when any such tank, receptacle, equipment or plumbing fixture is flooded or subject to pressure in excess of the operating pressure in the water piping. Where this possibility exists, an air gap or an approved back-flow prevention device should be provided.

(8) Piping exposed to possible freezing should be insulated and provided with electric heating cable.

(9) Where site water mains are looped, sectionalizing valves should be provided.

(10) On-site fire hydrants should be located so that, together with street fire hydrants, all parts of all buildings will be covered by hose streams from two hydrants, using not over 500 ft of hose

each. This coverage need be provided only for the outside of buildings provided with interior fire hose stations.

(a) Locate fire hydrants around the buildings, preferably no closer than 20 ft nor more than 50 ft from the building.
(b) Locate fire hydrants so that they are accessible to the fire department from an adjacent roadway. Review this with the fire department and architect.
(c) The minimum size branch to fire hydrants should be 6 in and provided with a curb valve.

(11) Size piping on building loads (500 gpm minimum for fire standpipe system and the maximum estimated sprinkler flow) with an allowance of 1,000 gpm for fire hydrants.
(12) Ample provision must be made in the pipe sizing for anticipated future building loads.
(13) Services should be provided with approved type meters; and where there are two services to the site, a check valve must be provided in each service. (Compound-type meters should be provided with a straight pipe ahead of them equal to at least eight pipe diameters.)
(14) Water meters, back-flow preventers, etc., should be located in pits or inside the building as required by the local water department or water company. (Obtain all details and requirements for the meter installation from the local water department or the local water company.)
(15) Domestic and fire water services should normally be split outside the building.
(16) Underground piping must have the minimum cover used in the area of the project. Requirements can be obtained from the local water department or water company.

2. Domestic water supply systems.

a. Principles of design.

(1) Building requirements (both initial and ultimate) should be obtained from the building project engineer.

(a) Ample provision should be made in the piping and in the equipment where practical, for anticipated future building expansion.

(2) Where no site fire mains with fire hydrants are being provided, the required site fire hydrants should be connected to the site domestic water main and normally have two 2-1/2 in. connections and a pumper connection.

(3) Where the basic fire hydrants are connected to a site fire main, at least one fire hydrant should be provided on the site domestic water main adjacent to a siamese on the site fire main to allow the fire department to boost the pressure in the site fire main.

(4) Where on-site storage must be provided, it should be in steel or concrete tanks, preferably above ground. Do not use concrete tanks below ground without specific approval from the water department or water company, or where no public supply is involved, from the Department of Health.

(a) Refer to NFPA Standard No. 22, Water Tanks, for construction of tanks.

(5) Booster pumping stations, when required, may be located in the building (designed under the building system) or in a separate building designed under the site work.

(a) The site designer should work closely with the building project engineer as to the capacity and type of equipment involved.
(b) Refer to Section 9. Building Domestic Water Supply Systems for pipe sizing and further data on equipment.

(6) All building domestic water services and fire hydrant branches should be provided with curb valves.
(7) The maximum allowable velocity in all of the water piping is 8 fps, except for the maximum velocity in pump suction connections, which is approximately 5 fps.

b. Piping.

(1) Unless otherwise noted, underground piping 3 in and larger should be cement-lined class 150 cast iron or thickness class 2 ductile iron bell and spigot water pipe and fittings.
(2) For underground piping beyond 8 ft outside the building walls, class 150 asbestos cement pressure piping may be considered if its use is common practice in the area.
(3) Underground piping 2 in and smaller should be type K copper tubing with cast brass flared type fittings.
(4) Underground valves shall be AWWA IBBM double-disc gate valves with curb boxes.
(5) Friction in cement-lined cast-iron, cement-lined ductile-iron, and asbestos cement piping should be calculated on the basis C = 140.
(6) For irrigation systems, refer to Section D in Section 9.
(7) For swimming pools, refer to Section E in Section 9.

(8) For decorative pools and fountains, refer to Section 9.

3. Fire-protection systems.

a. Principles of design.

(1) Obtain building requirements (both initial and ultimate) from the building project engineer. Make ample provision in the piping for anticipated future building expansion.
(2) Design fire-protection systems in accordance with the requirements of the fire chief or the fire marshall, and the owner's insurance underwriters. (Systems should be tentatively designed on the basis of NFPA standards and good engineering judgement and submitted to the authorities for approval and/or comment.)
(3) For fire-protection standards, refer to NFPA Standards 13, Sprinklers; 14, Standpipe and Hose Systems; 20, Fire Pumps; 22, Water Tanks; and 24, Outside Protection.
(4) Materials and equipment must be approved by the underwriters.
(5) Where separate site fire mains are provided, they should be looped around the buildings with adequate sectionalizing valves.
(6) Provide separate site fire mains with a siamese adjacent to a fire hydrant connected to the site domestic water main or the street main to allow the fire department to boost the pressure in the site fire main.
(7) Provide fire hydrants on the site fire main. If the fire main is fed from a fire pump or siamese, the fire hydrants normally have only two 2-1/2-in connections.
(8) For on-site storage, steel or concrete tanks, preferably above ground, must be

provided. Elevated steel tanks on a tower may be required. (Refer to NFPA Standard 22, Water Tanks, for construction of tanks.)

(9) Fire pumping stations, when required, may be located in the building (designed under the building system) or in a separate building designed under the site work.

  (a) The site designer should work closely with the building project engineer concerning the capacity and type of equipment involved.
  (b) Refer to Section 9 for further data on equipment.
  (c) A jockey pump of generous capacity should always be provided.

(10) The system can have fixed connections from either the public (domestic) water supply or another water source such as a pond or lake, but not both.

(11) When the system is connected to the public water supply and a pond or lake exists on the site (which the fire department might use as suction for their pumpers pumping into a siamese), an approved back-flow preventer should be provided in the public water supply connection.

b. Piping.

(1) Unless otherwise noted, underground piping must be cement-lined class 150 cast-iron or thickness class 2 ductile-iron bell and spigot water pipe with 250-psi cast-iron fittings.

(2) For underground piping beyond 8 ft outside the building walls, class 150 asbestos cement pressure pipe with 250-psi cast-iron fittings may be considered, if its use is common practice in the area and it is approved by the underwriters.

(3) Underground cast-iron and asbestos cement piping pressured from a fire pump or a siamese should be class 250.

(4) Inside piping should be standard-weight steel pipe with standard weight cast-iron threaded or flanged fittings, or Victaulic malleable iron fittings and couplings. Where fire pump discharge pressure may be over 175 psi, use extra heavy fittings. (Jockey pump piping and piping around a fire pump subject to water flow during periodic tests should be galvanized.)

(5) Underground valves should be post-indicator type or curb type, as required.

(6) For inside valves, refer to Sections H.2 and I.2, Section 9.

(7) Friction in steel-fire line piping can be calculated on the basis of C = 120.

(8) Friction in cement-lined cast-iron, cement-lined ductile-iron, and asbestos cement piping can be calculated on the basis C = 140.

D. Gas systems.

1. Principles of design.

a. Obtain building requirements (both initial and ultimate) from the building project engineer. (Make ample provision in the piping for anticipated future building expansion.)

b. Gas is usually supplied from the utility company's street mains. When no street mains are available, supply must be from a liquified petroleum (LP) "bottled" installation.

c. Contact the local utility company for the availability of street (natural) gas in the area. Obtain drawings showing locations and sizes of gas mains in the area. Obtain pressure, Btu/cu ft content of the available gas, and requirements regarding metering. (Gas services (natural gas) should be extended from the utility company's mains to all buildings requiring same.)

d. When street gas is not available, LP bottled gas must be provided.

   (1) Contact the local LP gas supplier concerning the size of the tank recommended for the load involved and find out whether an underground or an aboveground installation is recommended and also who supplies the tank.
   (2) Review with the architect both the tank location and the determination of an underground or an aboveground installation.

e. Design gas systems in accordance with the utility company's requirements, the requirements of applicable codes, and the requirements of NFPA Standards 54, National Fuel Gas Code, or 58, Liquefied Petroleum Gases, Storage and Handling.
f. Gas service is usually low pressure (1/2 psi or less). However, it may sometimes be medium or high pressure and require reduction. Check with the utility company and the code about whether it is to be reduced inside or outside the building.
g. Gas is usually metered inside the building, but it is sometimes metered on the site. Check with the utility company.
h. Gas piping from the street main to the property line is normally provided by the utility company. Gas piping inside the property line is usually provided by the plumber. (Obtain from the utility company the allowable pressure drop from the street main to the meter, from the street main to the regulator, from the regulator to the meter, whichever is applicable.)

2. Piping.

   a. Design piping in accordance with the requirements of the local code and the utility company (determine the extent of cathodic protection required).

b. Piping must be made of materials recommended by the utility company.

   (1) Normally piping is black steel pipe with welded or "dresser" fittings.
   (2) Often it is coated "mill wrapped."
   (3) Often it is provided with cathodic protection.

c. Piping from the street main to the meters is usually sized by the utility company. You are responsible for sizing the piping from the meter to the point of use.

   (1) When the meter is on the individual building service, the building project engineer will size the piping into the building.
   (2) When the meter is on the site service feeding a site distribution system, the site distribution system piping will be sized cooperatively by the building project engineer and the site engineer, as the total pressure drop allowed between the meter and the most distant point of use can be only 0.3 in of water.

d. Curb valves should be lubricated plug valves with curb boxes.

Section **2**

Plumbing Calculation Forms

Designers can save time and money by using calculation forms to help in plumbing design. The reason for this is that the forms help prevent the designer from overlooking important items. The forms in this chapter have been proven on many plumbing design assignments for a variety of structures.

Starting with electrical connection data, the designer is provided forms covering plumbing connections for HVAC equipment, intensity duration for storm water drainage, storm water drainage pipe sizing, domestic water system data, direct feed street pressure system, pump calculations, riser sizing, main sizing, hot water recirculation system sizing, gas riser sizing, gas main sizing, laboratory gas sizing, air sizing, vacuum sizing, oxygen sizing, and nitrous oxide sizing.

Using these charts will help the designer whenever the design has elements covered by the data provided. A neat set of these forms is an excellent backup for any plumbing system design.

FORM A

ELECTRICAL CONNECTION DATA

SHEET ______ OF ______

DATE ______________

REVISED ______________

JOB ______________ CURRENT CHAR ______________ V ______ PH ______ WIRE ______ CY.

ISSUED TO ______________ V ______ PH ______ WIRE ______ CY.

MOTORS ______ HP OR LESS: ______ VOLTS ______ PHASE

MOTORS ______ HP OR OVER: ______ VOLTS ______ PHASE

EQUIPT. ______ KW OR LESS: ______ VOLTS ______ PHASE

EQUIPT. ______ KW OR OVER: ______ VOLTS ______ PHASE

MOTORS ______ HP OR LESS: FULL INRUSH STARTING

MOTORS ______ HP OR OVER: REDUCED INRUSH STARTING

NO. OF OVERLOAD DEVICES FOR 3 PHASE MOTORS ☐ 2 ☐ 3

STARTERS:

TYPE	TYPE OF DISCONNECT	TYPE OF STARTING

SYSTEM NO.	EQUIPMENT AND LOCATION	HP OR WATTS	STARTER TYPE	INTERLOCK SYST. NO.	SMOKE DET. HEAD LOC.	ACTUATING & REMOTE CONT. DEVICES & LOCATIONS	REMARKS

FORM B

PLUMBING CONNECTIONS REQ'D FOR H.V.A.C. EQUIPT.

JOB ______________________ ISSUED TO ______________ DATE __________

RETURN TO PLBG. DESIGNER AS EARLY AS POSSIBLE ACCOMPANIED WITH MARKED UP PRINTS.

NO.	EQUIPMENT	LOCATION	C.W.				GAS			DRAIN		
			USE	PR	gpm	SIZE	PR	BTU HR	SIZE	USE	gpm	SIZE
	ALARMS											

FORM C Sheet 1

JOB ______________________ SHEET NO. ______ OF ______ SHEETS

SYSTEM STORM WATER DRAINAGE BY ____________ DATE ____________

(SHEET 1 OF 3)

RATE OF RAINFALL =
YEAR PERIOD

INTENSITY DURATION = i

C.B.Nº	DISTANCE WATER TRAVELS	CHANGE IN GRADE ELEVATIONS Δh	GRADE SLOPE	INLET CONCENT. TIME (min)	i RAINFALL INTEN. (in/hr)	REMARKS

FORM C Sheet 2

JOB ______________________ SHEET NO. ______ OF ______ SHEETS

SYSTEM STORM WATER DRAINAGE (SHEET 2 OF 3) BY ____________ DATE ____________

Q = A.C. i

C.B. Nº	⊞ PAVED	100% (C)* ADJ. PAVED	⊞ GRASS	35% (C)* ADJ. GRASS	TOTAL ADJ. ⊞	A.C. ⊞ 43, 560 = A.C. IN ACRES	i	Q = A.C. i = CFS

*VARIABLE

FORM C Sheet 3

JOB ______________________ SHEET NO. ______ OF ______ SHEETS

SYSTEM SYSTEM WATER DRAINAGE BY ____________ DATE ____________

(SHEET 3 OF 3)

PIPE SIZE

SIZE																												
INLET EL. OF ENTER. MH/CB																												
DROP																												
SLOPE %																												
FT. OF RUN																												
TO MH OR CB Nº																												
OUTLET ELEV.																												
TOTAL CFS																												
CFS. C.B.																												
C.B. OR MH Nº																												

FORM D

JOB ______________________ DATE ______________________ BY ______________________

STORM WATER DRAINAGE SIZING							
LDR. NO.	LDR ϕ	PITCH	SIZE	MAIN	PITCH	SIZE	REMARKS

FORM E

JOB ______________________ DATE ______________ BY ____________

STORM WATER SIZING SHEET

COMM. TO MAIN #	Φ BRANCH CONN.	TOTAL Φ BRANCH	BRANCH SIZE	TOTAL Φ MAIN	MAIN SIZE

FORM F

JOB ______________________________ DATE: ____________________ BY ____________________

DRAINAGE, SIZING

WING	STACK OR FIXTURES	FIXTURE UNITS BR.	TOTAL F.U. MAIN	SIZE	REMARKS

FORM G Sheet 1

JOB ______________________ SHEET NO. 1 OF ______ SHEETS

SYSTEM ______________________ BY __________ DATE __________

DOMESTIC WATER SYSTEM

AVAILABLE: ______ * PSI AT ELEV. ______ IN ____________ STREET

Total St. Pr. FU ________

St. Pr. GPM ________

Pump GPM ________

HVAC GPM ________

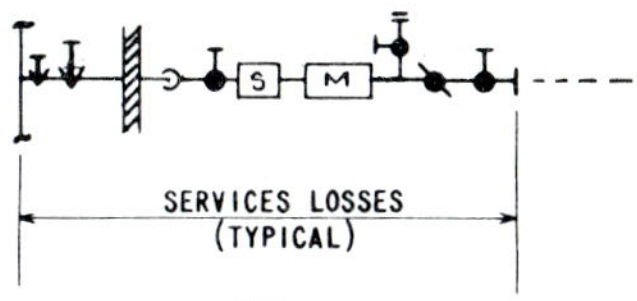

Total GPM ________

Size ________

Frict. Loss/100' ________

Allowance for future drop in St. Pr. __________ PSI

(1) * Adjusted Available St. Pr. __________ PSI x 2.31 = __________ ft.

(2) SERVICE LOSSES

No.	Size	Item				
		Gate Valves	@ ______	=	__________	ft.
		Check Valves	@ ______	=	__________	ft.
		90° Els (Misc Ftgs)	@ ______	=	__________	ft.
			Run		__________	ft.
			Total Footage	=	__________	ft.

Total Footage x Loss/100' = Fr Loss in Ft.

__________ x __________ = __________ ft.

______ " Strainer Loss @ ______ PSI x 2.31 = __________ ft.

______ " Meter Loss @ ______ PSI x 2.31 = __________ ft.

Total Friction Loss For Service [ft.] (2)

FORM G Sheet 2

JOB________________ SHEET NO. 2 OF ____ SHEETS

SYSTEM________________ BY________ DATE________

DIRECT FEED STREET PRESSURE SYSTEM

(3) LIFT

Elevation of top most fixture ________ ft.

Minus Elevation of St. Main - ________ ft.

LIFT [ ] (3)

(4) HEAD

Pressure Reqd. at top most fixture ____ PSI x 2.31 ________ ft.

Service Losses (2) ________ ft.

Lift (3) ________ ft.

Reqd. Head (Less Fr.) [ ] (4)

Adjusted Avail. St. Press. (1) ________ ft.

Minus Reqd. Head (4) - ________ ft.

Pressure Remaining For Friction [ft.]

CONCLUSIONS:

FORM G Sheet 3

JOB ________________ SHEET NO. 3 OF ____ SHEETS

SYSTEM ________________ BY ________ DATE ________

________ PUMP CALCULATION

(3) LIFT

Elevation one (1) ft. above house Tk.
Elevation of top most fixture } ________ ft.
Minus Elev. of Pump Room floor ________ ft.

LIFT [ft.] (3)

(4) Pump Suction Losses (After Service)*

No.	Size	Items			
		Gate Valves	@ ________	= ________	ft.
		Check Valves	@ ________	= ________	ft.
		90° Els (Misc Ftgs)	@ ________	= ________	ft.
			Run	________	ft.

Total Footage = ________ ft.

Total Footage x Fr. Loss/100' = Fr. Loss in Suction

________ x ________ = [ft.] (4)

(5) Pressure Available At Pump

Adjusted St. Pr. (1) ________ ft.
Gain to Pump Room Fl. Elev. ________ ft.
________ ft.

Minus { Service Losses (2) ________ ft.
Suction Losses (4) ________ ft.
________ ft. ← ________ ft.

Available Pressure At Pump [ft.] (5)

* NOTE: For complicated suction piping, additional calculations will be required. Use main sizing sheets.

FORM G Sheet 4

JOB ______________ SHEET NO. 4 ___ OF ___ SHEETS

SYSTEM ______________ BY ______ DATE ______

(6) Pump Discharge (System) Losses

No.	Size	Item			
		Gate Valves	@ ______	= ______	ft.
		Check Valves	@ ______	= ______	ft.
		90° Els (Misc Ftgs)	@ ______	= ______	

Total Footage = ______ ft.

Total Footage x Fr. Loss/100' = Total Fr. Loss in Discharge

______ x ______ = ______ ft. (6)

HEAD

Pressure Reqd. at top of System ____ PSI x 2.31 = ______ ft.

Lift (3) ______

Pump Discharge Losses (6) ______

Total ______

Minus Adjusted Avail. Pressure At Pump (5) ______

REQUIRED HEAD ______

CONCLUSIONS:

FORM H

RISER SIZING – STACK Nº

FLOOR	PRESSURE		F.U. DRAIN	COLD WATER								HOT WATER							
				F.U. BR.	G.P.M. BR.	SIZE BR.	F.U. RISER	G.P.M. RISER	SIZE RISER	FR. FT/100	FR. LOSS	F.U. BR.	G.P.M. BR.	SIZE BR.	F.U. RISER	G.P.M. RISER	SIZE RISER	FR. FT/100	FR. LOSS

HEIGHT OF SOIL SOIL STACK F.U.
HEIGHT OF VENT VENT STACK F.U.

FORM I

MAIN SIZING - SHEET №																			
				COLD WATER								HOT WATER							
NO.	RUN		F.V. DRAIN	F.V. BR.	G.P.M. BR.	SIZE BR.	F.V. MAIN	G.P.M. MAIN	SIZE MAIN	FR. FT/100	FR. LOSS	F.V. BR.	G.P.M. BR.	SIZE BR.	F.V. MAIN	G.P.M. MAIN	SIZE MAIN	FR. FT/100	FR. LOSS

FORM J

HOT WATER RECIRCULATION SYSTEM SIZING FORM

CAL. GPM CIR PUMP CAPACITY													CAL. FT HEAD OF PUMP				
SUPPLY						RETURN						TOTALS		RETURN			
SECT	SUB SECT	DIA	MEAS L	HT. LOSS RATE	HEAT LOSS	SECT	SUB SECT	DIA	MEAS L	HT. LOSS RATE	HEAT LOSS	TOT. HEAT LOSS	Q = ** H.L. / 5000	EQUIV LGTH. 1½xL	FR/ 100 ft	FR LOSS	TOT GOV FR*
TOTALS:–																	

*Total friction loss (ft) from longest or governing run = pump head.

**Total of column = gpm capacity of pump.

FORM K

JOB ______________________ DATE ______________ BY ______________

GAS RISER SIZING										
FLOOR		OUTLETS BR	C.F.H. BR	SIZE BR	OUTLETS RISER	DEMAND FACTOR	C.F.H. RISER	SIZE RISER	PRESS LOSS in/100	PRESS. LOSS

FORM L

JOB ______________________ DATE ______________________ BY ______________________

GAS MAIN SIZING										
NO.	EQUIV. RUN	OUTLETS BR.	C.F.H. BR.	SIZE BR.	EQUIP/ OUTLETS MAIN	DEMAND FACTOR	C.F.H. MAIN	SIZE MAIN	PRESS. LOSS in/100	PRESS. LOSS

FORM M

JOB ______________________ DATE ______________ BY ______________

LABORATORY GAS SIZING SHEET NO. ______________________

LOCATION ______________________

TOTAL LOSS IN MAIN BEYOND BRANCH CONN ______________________

________ CFH/BUNSEN BURNER

NO.	ACTUAL RUN	FITTINGS	EQUIV. RUN	CFH BRANCH	SIZE BRANCH	TOTAL CFH	FACTOR	ADJ. CFH	SIZE	LOSS	TOTAL LOSS	

FORM N

JOB ______________________ DATE ______________ BY ______________

AIR SIZING – SHEET NO.

NO.	RUN	FITTING	EQUIV. RUN	CONN. CFM BRANCH	SIZE BRANCH	TOTAL MAIN	FACTOR	ADJ. CFM	SIZE	FR. psi/100 ft	FR. LOSS	NO.	RUN	FITTING	EQUIV. RUN	CONN. CFM BRANCH	SIZE BRANCH	TOTAL MAIN	FACTOR	ADJ. CFM	SIZE	FR. psi/100 ft	FR. LOSS

FORM O

JOB ____________________ DATE ______________ BY ____________________

VACUUM SIZING SHEET NO.

CONN NO.	RUN	FITTINGS	EQUIV RUN	1 FACTOR ___%				2 FACTOR ___%				3 FACTOR ___%				ADJ BRANCH CFM				TOTAL ADJ. CFM (TOTAL FROM 1,2,3)	SIZE	FR. in Hg/100 ft	FRICTION LOSS
				BRANCH CFM	BRANCH SIZE	TOTAL CFM	ADJ. CFM	BRANCH CFM	BRANCH SIZE	TOTAL CFM	ADJ. CFM	BRANCH CFM	BRANCH SIZE	TOTAL CFM	ADJ. CFM	NO. @___%	NO. @___%	NO. @___%	TOTAL ADJ. CFM				

FORM P

JOB ______________________ DATE __________ BY __________

OXYGEN SIZING SHEET NO.

NO.	RUN	TOTAL RUN	OUTLETS	TOTAL OUTLETS	ADJ. LTR/MIN	SIZE	NO.	RUN	TOTAL RUN	OUTLETS	TOTAL OUTLETS	ADJ. LTR/MIN	SIZE

FORM Q

JOB ______________________ DATE ____________ BY ______________

NITROUS OXIDE SIZING SHEET NO.

NO.	RUN	TOTAL RUN	OUTLETS	TOTAL OUTLETS	ADJ. LTR/MIN	SIZE	NO.	RUN	TOTAL RUN	OUTLETS	TOTAL OUTLETS	ADJ. LTR/MIN	SIZE

Section **3**

Sizing the Water Supply System

All buildings in which human beings spend time to work or live require a water supply of some type. The most important design objective in sizing the water supply system is to furnish a sufficient amount of potable water at proper pressure and flow rate for normal fixture operation.

Sizing of water supply systems for buildings is an important design procedure which is vital in protecting the health of the occupants. Hence, pipe sizing is regulated by various plumbing codes.

Other important objectives in the design of building water supply systems are: (1) achieve economical sizing of the piping, while eliminating overdesign; (2) provide against potential supply failure from gradual reduction of the pipe bore with the passage of time; (3) avoid erosion-corrosion effects and potential pipe failure or leakage conditions because of corrosive characteristics of the water; and (4) eliminate water hammer damage and objectionable noise in piping from excessive design velocities.

DESIGN OBJECTIVES

The most important design objective in sizing the water supply system is the satisfactory supply of potable water to all fixtures, at all times, and at proper pressure and flow rate for normal fixture operation. This may be achieved only if adequate sizes of pipes and appurtenances are provided. The sizes established must be large enough to prevent occurrence of negative pressures in any part of the system during periods of peak demand in order to avoid the hazard of water supply contamination due to backflow and back siphonage from potential sources of pollution. Hence, the sizing of building water supply systems is a matter of vital concern in protecting health and must be regulated by codes.

Other important objectives in the design of water supply systems are: (1) to achieve economical sizing of piping and eliminate overdesign; (2) to provide against potential supply failure due to gradual reduction of pipe bore with the passing of time, such as may result from deposits of corrosion or hard water scale in the piping; (3) to avoid erosion-corrosion effects and potential pipe failure or leakage conditions owing to corrosive characteristics of the water and/or to excessive design velocities of flow; and (4) to eliminate water hammer damage and objectionable whistling noise effects in the piping due to excessive design velocities of flow.

MINIMUM AVAILABLE PRESSURE

Design Pressures for System The water supply system should be designed in accordance with the minimum pressure available at the public water main, or other source of water supply pressure, and the minimum pressure required at all times at water outlets of the system.

Adequate Source of Water Supply Pressure Where the pressure available at the public main is insufficient to maintain the minimum required at the highest water outlet of the system, a pressure booster pump system, approved as to capacity and reliability, or an automatically controlled water supply tank of either the hydropneumatic pressure type or the elevated, gravity type should be provided.

Pressure Booster Pump System For a pressure booster pump system, the minimum pressure at the pump outlet designed to be maintained at all times should be considered the minimum pressure available at the source.

Automatically Controlled Water Supply Tanks Automatically controlled water supply tanks should be designed and installed so as to furnish the required minimum pressures. Where a pressure tank is provided, the minimum pressure at which it is designed to operate should be considered the minimum available at the source. Where a gravity tank is provided, the minimum pressure at the source should be considered as being the static pressure owing to the elevation of the low waterline at which the tank is designed to operate under peak demand conditions.

Water Outlets As a general rule, the minimum pressure required at ordinary faucets of plumbing fixtures should be deemed to be 8 psi (55 kPa). However, at direct supply-connected flush valves (Flushometers), the minimum should be 25 psi (172 kPa) for blow-out-type water closets and 15 psi (103 kPa) for other types of fixtures. At electrically operated supply valves of equipment, where higher than ordinary pressure frequently is recommended, the minimum should be that recommended by the manufacturer for satisfactory equipment performance.

DEMAND AT INDIVIDUAL WATER OUTLETS

Ordinary Pipe Outlets The demand at ordinary pipe outlets, flowing wide open at maximum rate of discharge, may be calculated approximately based on the flow pressure in the supply pipe and the diameter of the outlet. This may be done by applying the formula for the practical jet discharge from pipe outlets, which is as follows:

$$q_d = 20\,(d_o)^2\sqrt{p_m}$$

where q_d = actual quantity rate of discharge from pipe outlet, gpm
d_o = diameter of outlet, in
p_m = pressure measured in supply pipe during flow, psi

Electrically Operated Supply Valves Equipment having electrically operated supply valves varies considerably in demand requirements depending upon the type used and its performance characteristics. Information about demand may be obtained from data furnished by manufacturers of such equipment. In the absence of adequate information, the demand may be calculated based on the flow pressure required and the outlet diameter, just as for any ordinary pipe outlet. Usually, electrically operated supply valves are kept wide open for maximum flow rate.

Common Plumbing Fixtures Demand rates which have become generally accepted as normal, suitable rates of flow from individual water outlets at various plumbing fixtures and hose connections are given in Table 3-1. The demands shown for faucets which users may adjust manually according to their needs are not the maximum flow rates that may be obtained in each case. Rather, they are demands which are considered suitable for fixture usage without causing excessive splashing, a factor which is related to the shape and depth of the fixture.

Table 3-1
DEMAND AT INDIVIDUAL WATER OUTLETS

Type of outlet	Demand	
	gpm	L/s
Ordinary lavatory faucet	2.0	0.126
Self-closing lavatory faucet	2.5	0.158
Sink faucet, ⅜" (9.52 mm) or ½" (12.7 mm)	4.5	0.284
Sink faucet, ¾" (19 mm)	6.0	0.378
Bath faucet, ½" (12.7 mm)	5.0	0.315
Shower head, ½" (12.7 mm)	5.0	0.315
Laundry faucet, ½" (12.7 mm)	5.0	0.315
Ball cock in water closet flush tank	3.0	0.189
1" (25.4 mm) flush valve [25 psi (172 kPa) flow pressure]	35.0	2.210
1" (25.4 mm) flush valve [15 psi (103 kPa) flow pressure]	27.0	1.703
¾" (19.0 mm) flush valve [15 psi (103 kPa) flow pressure]	15.0	0.946
Drinking fountain jet	0.75	0.047
Dishwashing machine (domestic)	4.0	0.252
Laundry machine [8 lb (3.6 kg) or 16 lb (7.3 kg)]	4.0	0.252
Aspirator (operating room or laboratory)	2.5	0.158
Hose bib or sill cock, ½" (12.7 mm)	5.0	0.315

Control of Excessive Flow Rates at Water Outlets To avoid excessive flow rates and splashing at fixtures where the available pressure is considerably higher than the minimum required, it is generally necessary to provide some means of reducing the maximum flow rate to match normal demand. This is especially needed at fixtures located on the lower floors of high and tower-type buildings, for at such locations the pressure existing in water supply systems is many times the minimum required at water outlets.

Where the available pressure at water outlets is more than twice the minimum pressure required for satisfactory supply, it is recommended that means to control the rate of flow should be provided in the fixture supply pipe. For this purpose, individual regulating valves, variable orifice control devices, or fixed orifices may be provided.

Flush valves (Flushometers) are usually equipped with adjustment screws for regulating valve operation and are generally provided with a throttling valve at the valve inlet so as to reduce the available flow pressure to 25 or 15 psi (172 or 103 kPa), as may be required for satisfactory function of water closets, urinals, bedpan washers, flushing-rim slop sinks, or dishwashing machines.

For faucets and flush tank ball cocks, a throttling valve or flow control orifice may be installed in the fixture supply pipe to reduce the maximum flow rate to match normal demand. To control faucet flow, throttling valves should be adjusted so that flow matches demand when the faucet is wide open. Flow control orifices should be selected in accordance with desired demand and anticipated supply pressures.

Flow Control for Conservation of Water and Energy As a water conservation measure, flow rates at certain faucet outlets may be reduced slightly below those shown in Table 3-1, without causing any noticeably adverse effect in usage. Fixtures at which this may be applied and the minimum flow rates recommended are as follows: lavatory faucets in private bathrooms, 1.5 gpm (0.095 L/s); shower heads in private bathrooms, 2 gpm (0.126 L/s); and sink faucets at domestic kitchen sinks, 3 gpm (0.189 L/s).

As an energy conservation measure, the maximum flow rate for hot water lavatory faucets in restrooms to which the general public has access is recommended to be limited by design to ½ gpm (0.032 L/s), and the outlet temperature limited to 110°F (43.4°C). This requirement may be found in state energy conservation construction codes. In view of such requirement as to maximum flow rate for the hot water lavatory faucet in restrooms for the general public, it is recommended that the same limitation be observed for the cold water lavatory faucet.

ESTIMATING DEMAND

Need for a General Method for Estimating Demand The objectives in designing water supply systems for buildings are to ensure adequate water supply to all fixtures at all times and to achieve economical sizing of piping. To do this on a rational basis, it is necessary to estimate as accurately as possible the probable maximum rate of flow or the demand for which provision should be made in every portion of the system, including the water service, main supply lines, risers, and main branches.

Demand in building water supply systems cannot be determined exactly. Although some water outlets, such as those for watering gardens, washing sidewalks, irrigating lawns, and air-conditioning apparatus, impose continuous demand during periods of peak demand, the fact must be recognized that most plumbing fixtures in buildings are used intermittently, and the probability of simultaneous use of such fixtures cannot be definitely established. In addition, each type of plumbing fixture imposes its own singular loading effect on the system. This may be attributed to (1) average rate of supply required by a fixture for satisfactory service, (2) duration of fixture use, and (3) frequency of fixture use. Nevertheless, the demand imposed on the building water supply system by intermittently used fixtures is clearly related to the number, type, and probable simultaneous use of the fixtures to be supplied.

Many methods for estimating demand have been applied in the past. Most have been found to be unsatisfactory for one reason or another. Some resulted in considerable underestimates of demand; hence, undersizing of piping and consequent failure of supply or inadequate service to fixtures. By contrast, some produced extreme overestimates and uneconomical oversizing of systems. Still other methods were found unsuitable because they were adapted just to groups of fixtures of the same kind and were not applicable to large numbers of different types of fixtures. Some recent methods have been developed by plumbing engineering designers for specific building occupancy classifications, and they have been reportedly effective in achieving greater economy in such cases. However, such methods are appropriate only within their proved limits. They are not methods recommended for general application.

For a method to be generally acceptable for estimating the demand of a building water supply system, it must meet three basic requirements. First, it must produce estimates greater than the average demand for all fixtures on the system during periods of heaviest demand; otherwise failure will occur in the supply to some fixtures during maximum demand periods. Second, the method must yield reasonably accurate estimates of peak demand so as to avoid oversizing of piping and uneconomical waste. Third, the method must be adaptable for estimating the demand of groups of like fixtures as well as of different kinds of fixtures and building occupancy classifications.

Standard Method A generally accepted standard method for estimating the demand of building water supply systems has evolved and become recognized as meeting the above-mentioned three basic requirements. In 1923, a method of weighting plumbing fixtures in accordance with their water supply load producing effects was proposed by Roy B. Hunter of the National Bureau of Standards, U.S. Department of Commerce. This method was presented in the publication, "Recommended Minimum Requirements for Plumbing," Department of Commerce Building and Housing Series, BH13, 1924. After study of the application of the proposed method in the design of federal buildings over a period of years, the method was revised slightly and presented by the National Bureau of Standards in a 1940 publication, Building Materials and Structures Report, BMS65, "Methods of Estimating Loads in Plumbing Systems," by Roy B. Hunter.

In 1968, appropriate modifications were made in the method for modern plumbing fixtures as the result of research studies reported by Elmer Jones of the Agricultural Research Service, U.S. Department of Agriculture. These research studies were made on small housing developments and reflected the evident need posed by modern plumbing fixtures for changes in the demand values, in gallons per minute, corresponding to load values of from 1 to 60 water supply fixture units.

Up to the present, this method has filled the need for a reliable, rational way to estimate the demand in water supply systems for modern buildings of all types of occupancy and has become recognized as a generally accepted method.

Load Values (WSFUs) Assigned to Fixtures Demand in building water supply systems cannot be determined exactly. The demand imposed on a system by intermittently used fixtures is related to the number, type, and probable simultaneous use of the fixtures to be supplied.

In the standard method, fixtures using water intermittently under several conditions of service are assigned specific load values in terms of water supply fixture units. The water supply fixture unit (WSFU) is a factor so chosen that the load producing effects of different kinds of fixtures and their conditions of service can be expressed as multiples of that factor.

Values assigned to different kinds of fixtures are given in Table 3-14. For fixtures having both hot and cold water supplies, the values for separate hot and cold water demands should be taken as being three-fourths of the total value assigned to the fixture in each case. For example, the value assigned to a kitchen sink in a dwelling unit is 2 WSFU, while the separate demands on the hot and cold water piping thereto should be taken as being 1.5 WSFU.

Demand Corresponding to Fixture Load To determine the demand in gallons per minute or liters per second corresponding to any given load in water supply fixture units, reference may be made to two charts shown in Fig. 3-4 and also specifically mentioned later in this chapter under the subject, "Standard Code Regulations."

As an aid in this regard, tabulated values for the demand in gallons per minute or liters per second corresponding to given loads in water supply fixture units have been determined from the two curves and arranged conveniently in Table 3-2. In this way, easy conversion may be made from terms of water supply fixture units to gallons per minute or liters per second of flow, or vice versa. Intermediate values may be interpolated for loads between those shown in the table.

Table 3-2
TABLE FOR ESTIMATING DEMAND

Supply systems predominantly for flush tanks			Supply systems predominantly for Flushometer valves		
Load	Demand		Load	Demand	
Water supply fixture units (WSFU)	gpm	L/s	Water supply fixture units (WSFU)	gpm	L/s
1	3.0	0.19			
2	5.0	0.32			
3	6.5	0.41			
4	8.0	0.51			
5	9.4	0.59	5	15.0	0.95
6	10.7	0.68	6	17.4	1.10
7	11.8	0.74	7	19.8	1.25
8	12.8	0.81	8	22.2	1.40
9	13.7	0.86	9	24.6	1.55
10	14.6	0.92	10	27.0	1.70
12	16.0	1.01	12	28.6	1.80
14	17.0	1.07	14	30.2	1.91
16	18.0	1.14	16	31.8	2.01
18	18.8	1.19	18	33.4	2.11
20	19.6	1.24	20	35.0	2.21

Table 3-2 (*Continued*)

Supply systems predominantly for flush tanks			Supply systems predominantly for Flushometer valves		
Load	Demand		Load	Demand	
Water supply fixture units (WSFU)	gpm	L/s	Water supply fixture units (WSFU)	gpm	L/s
25	21.5	1.36	25	38.0	2.40
30	23.3	1.47	30	42.0	2.65
35	24.9	1.57	35	44.0	2.78
40	26.3	1.66	40	46.0	2.90
45	27.7	1.76	45	48.0	3.03
50	29.1	1.84	50	50.0	3.15
60	32.0	2.02	60	54.0	3.41
70	35.0	2.21	70	58.0	3.66
80	38.0	2.40	80	61.2	3.86
90	41.0	2.59	90	64.3	4.06
100	43.5	2.74	100	67.5	4.26
120	48.0	3.03	120	73.0	4.61
140	52.5	3.31	140	77.0	4.86
160	57.0	3.60	160	81.0	5.11
180	61.0	3.85	180	85.5	5.39
200	65.0	4.10	200	90.0	5.68
250	75.0	4.73	250	101.0	6.37
300	85.0	5.36	300	108.0	6.81
400	105.0	6.62	400	127.0	8.01
500	124.0	7.82	500	143.0	9.02
750	170.0	10.73	750	177.0	11.17
1000	208.0	13.12	1000	208.0	13.12
1250	239.0	15.08	1250	239.0	15.08
1500	269.0	16.97	1500	269.0	16.97
2000	325.0	20.50	2000	325.0	20.50
2500	380.0	23.97	2500	380.0	23.97
3000	433.0	27.32	3000	433.0	27.32
4000	525.0	33.12	4000	525.0	33.12
5000	593.0	37.41	5000	593.0	37.41

In viewing the charts and table, it may be noted that the demand corresponding to a given number of water supply fixture units of load is generally much higher for a system in which water closets are flushed predominantly by means of direct supply-connected flush valves (Flushometers) than for a system in which they are flushed predominantly by means of flush tanks. The difference in demand between the two systems gradually diminishes as the total number of fixture units of load rises, until at 1000 water supply fixture units the demand for both types is the same, 208 gpm (13.12 L/s).

Where a part of the system does not supply any water closets, such as is the case with hot water supply piping and some cold water supply branches, the demand corresponding to a given number of water supply fixture units may be determined from the values given for systems in which water closets are flushed predominantly by means of flush tanks. The demands determined in such cases undoubtedly are appropriate in view of the average of values assigned to fixtures other than water closets having direct supply-connected flush valves (Flushometers).

Total Demand Including Continuous Flow To estimate the total demand in gallons per minute or liters per second in any given water supply pipe which supplies outlets at which the demand is intermittent and also outlets at which demand is continuous, the demand for outlets which pose continuous demand during peak periods should be calculated separately and added to the demand for plumbing fixtures used intermittently. Examples of outlets which impose continuous demand are those for watering gardens, washing sidewalks, irrigating lawns, and for air-conditioning or refrigeration apparatus.

PRELIMINARY INFORMATION

Information for Establishing Design Basis for Sizing Obtain all information necessary for establishing a proper basis for sizing the building water supply system. Properness of the basis for sizing is contingent upon accuracy and reliability of the information applied. Thus, such information should be obtained from responsible parties and appropriate authorities recognized as sources of the necessary information.

Materials for System The kind of piping materials to be installed in the system should be determined specifically. This is a matter of selection by the owner of the building or an authorized representative, who may be the architect, engineer, or contractor.

Characteristics of the Water Supply The corrosivity of a given water supply with respect to various kinds of piping materials, and of its scale-forming tendency, is information which most officials, architects, engineers, and contractors in a water district normally have at their fingertips as a result of years of experience. For anyone without such experience and knowledge, the significant characteristics which may be applied to indicate the water supply's corrosivity and scale-forming tendency are its pH value, CO_2 content, dissolved air content, carbonate hardness, Langelier index, and Ryznar index. The most appropriate source of such information is the local water authority having jurisdiction over the system supplying the water, or over the wells from which water is pumped from the underground water table.

Location and Size of Water Supply Source The location and size of the public water main, where available, should be obtained from the local water authority. Where a private water supply source, such as a private well system, is to be used, the location and size as designed for the premises should be applied.

Developed Length of System Information should be obtained regarding the developed length of the piping run from the source of water supply to the service control valve of the building, i.e., the developed length of the water service pipe as shown on site plans. Also, determine the developed length of the piping run from the service control valve to the highest and most remote water outlet on the system. This may be established by measurement of the piping run on plans of the system.

Pressure Data Relative to Source of Supply Data as to pressure available in the public water main should be obtained from the local authority having jurisdiction over the public system. Maximum and minimum pressures available in the public main at all times should be obtained from that authority, as it is the only one recognized as the source of accurate and reliable information on this subject. Where a private well water supply system is to be used, the maximum and minimum pressures at which it will be adjusted to operate may be applied as appropriate in such cases.

Elevations The relative elevations of the source of water supply and the highest water supply outlets to be supplied in the building must be determined. In the case of a public water main, the elevation of the point where the water service connection is to be made to the public main should be obtained from the local water authority. It has the most authoritative record of elevations of the various parts of the public system, and such elevations are generally referred to datum as the reference level, and related to curb levels established for streets. Elevation of the curb level directly in front of the building should be obtained from the building plans, as such information is required to be shown on the building site plans. Elevations of each floor at which fixtures are to be supplied may be determined also from the building plans.

Minimum Pressure Required at Highest Water Outlets Information regarding the minimum pressure required at the highest water outlets for adequate, normal flow conditions, consistent with satisfactory fixture usage and equipment function, may be deemed to be as follows: 8 psi (55 kPa) for all water supply outlets at common plumbing fixtures, except for direct supply flush valves at which the minimum required flow pressure should be considered to be 15 psi (103 kPa) when supplying floor-outlet-type water closets and urinals, and as 25 psi (172 kPa) when supplying wall-hung siphon-jet or blowout-type water closets. For other types of water supplied equipment, the minimum flow pressure required should be obtained from the manufacturer.

Provision of Necessary Information on Plans Information necessary for establishing a proper basis for designing sizes of water supply piping should be provided on plans of the water supply system when submitted to plumbing plan examiners for proposed installations. Provision of such information permits the examiner to check easily and efficiently the adequacy of sizes proposed for the various parts of the building water supply system.

LIMITATION OF FRICTION

Consideration of Friction in Design Friction in the building water supply system must be limited by design to the degree necessary so that the highest water outlets on the system will have available during periods of peak demand at least the minimum pressure required at such outlets for satisfactory water supply conditions.

Maximum Permissible Friction Loss The limit to which pressure loss due to friction may be permitted to occur in the main water lines and risers supplying the highest water outlets during peak demand periods is the amount of static pressure available in excess of the minimum pressure required at such outlets when no-flow conditions exist. This may be calculated as the difference between the static pressure existing at the highest water outlets during no-flow conditions, and the minimum pressure required at such outlets for satisfactory supply conditions.

Where supplied by direct pressure from a public main, to calculate the static pressure at the highest outlet, deduct from the certified minimum pressure available in the public main the amount of static pressure loss corresponding to the height at which the outlet is located above the public main; i.e., deduct 0.433 psi pressure for each foot (or 9.79 kPa/m) of rise in elevation from the public main to the highest outlet.

Where supplied under pressure from a gravity water supply tank located at an elevation above the highest water outlet, the static pressure at that outlet is calculated as being equal to 0.433 psi for each foot (or 9.79 kPa/m) difference in elevation between the outlet and the water line in the tank. In this case, the minimum static pressure at the outlet should be determined as that corresponding to the level of the low-water line at which the tank is designed to operate.

Where supplied under pressure from a pressure booster pump system, to calculate the static pressure at the highest outlet, deduct from the minimum pressure designed to be maintained at all times at the pump outlet the amount of static pressure loss corresponding to the height at which the highest outlet is located above the level of the pump outlet, i.e., deduct 0.433 psi pressure for each foot (or 9.79 kPa/m) of rise in elevation from the pump outlet to the highest water outlet.

Basic Design Circuit Of the highest water outlets on a system, the one at which the least available pressure will prevail during periods of peak demand is that outlet which is supplied through the longest run of piping extending from the pressure source of water supply, as a general rule. As pipe friction loss is directly proportional to length of piping, the most extreme run of piping from the pressure source of water supply to the highest outlets on the system should be designated as the "basic design circuit" (BDC) for sizing the main water lines and risers in accordance with the limit to which pressure loss due to friction may be permitted to occur therein.

In most systems, the BDC may be found to consist of the run of cold water supply piping extending from the pressure source of water supply to the domestic hot water heating unit, and the run of hot water supply piping therefrom to the highest and most remote hot water outlet on the system. However, in systems having flush valve (Flushometer) supplied water closets at the topmost floor, the BDC may be found to be the run of cold water supply piping extending from the pressure source of water supply to the highest and most remote flush valve on the system.

Friction Loss in Equipment Where a water meter, water filter, water softener, fish trap or strainer, or instantaneous or tankless hot water heating coil is provided in the basic design circuit, the friction loss corresponding to the peak demand through such equipment must be determined and included in pressure loss calculations. Manufacturers' charts and data sheets on their products provide such information generally, and should be used as a guide in selecting the most appropriate type and size of equipment to use with consideration for the limit to which pressure loss due to friction may be permitted to occur in the basic design circuit. The rated pressure loss through such equipment should be deducted from the friction loss limit so as to establish the amount of pressure which may be permitted to be dissipated by friction in pipe, valves, and fittings of the basic design circuit.

Pressure Loss in Displacement-Type Cold Water Meters The American Water Works Association standard for cold water meters of the displacement type is designated AWWA C700–71. It covers displacement meters known as nutating- or oscillating-piston or disk meters, which are practically positive in action.

The standard establishes maximum capacity or delivery classifications for each meter size as follows:

⅝ in—20 gpm (15.9 mm—1.26 L/s)

¾ in—30 gpm (19 mm—1.89 L/s)

1 in—50 gpm (25.4 mm—3.1 L/s)

1½ in—100 gpm (38.1 mm—6.3 L/s)

2 in—160 gpm (50 mm—10.1 L/s)

3 in—300 gpm (75 mm—18.9 L/s)

4 in—500 gpm (100 mm—31.5 L/s)

6 in—1000 gpm (150 mm—63 L/s)

The standard also establishes the maximum pressure loss corresponding to the stated maximum capacities as follows:

15 psi (103 kPa) for the ⅝-in (15.9-mm), ¾-in (19.0-mm), and 1-in (25.5-mm) meter sizes

20 psi (138 kPa) for the 1½-in (38.1-mm), 2-in (50-mm), 3-in (75-mm), 4-in (100-mm), and 6-in (150-mm) sizes

For estimating pressure loss in displacement-type cold water meters, Fig. 3-5 is provided. Pressure loss in meters for flow at less than maximum rates for any given size of meter may be estimated from Fig. 3-5.

Uniform Pipe Friction Loss To facilitate calculation of appropriate pipe sizes corresponding to the permissible friction loss in pipe, valves, and fittings, it is recommended that the basic design circuit be designed in accordance with the principle of uniform pipe friction loss throughout its length. In this way the friction limit for the piping run may be established in terms of pounds per square inch per 100 ft (or Pa/m) of piping length. The permissible uniform pipe friction loss is calculated by dividing the permissible friction loss in pipe, valves, and fittings by the total equivalent length of the basic design circuit, and multiplying by 100.

Equivalent Length of Piping The total equivalent length of piping is its developed length plus the equivalent pipe length corresponding to the frictional resistance of all fittings and valves in the piping run. When sizes of fittings are known or have been established by sizing based upon appropriate limitation of velocity, corresponding equivalent lengths may be determined directly from available tables.

As a general finding, experience indicates that the equivalent length to be allowed for fittings and valves as a result of such calculations is approximately 50 percent of the developed length of the basic design circuit in the case of copper water tube systems, and approximately 85 percent for standard threaded pipe systems.

ACCURATE PIPE FRICTION CHARTS FOR VARIOUS SERVICE CONDITIONS

Accurate pipe friction charts for various service conditions are presented herewith for each of the standard piping materials used for water supply systems in buildings. These charts were developed using the actual internal pipe diameters given in applicable standards for each of the different kinds of materials, rather than nominal diameters. All values calculated in developing the charts were made using the most authoritative pipe friction equations presented to date for application in accordance with the relative roughness of the internal pipe surface, i.e., roughness relative to pipe diameter.

Flow rates corresponding to any given uniform pipe friction loss may be determined readily for each nominal size of the kind of pipe selected for the system. The appropriate chart to apply in any given case depends upon the kind of piping to be used and the effect that the water to be conveyed will produce within the piping after extended service. On each of the charts, the particular equation applied and the internal pipe surface condition to which it relates have been stated. Accordingly, appropriate selection of the proper chart to use in determining flow rates through various sizes of a given kind of piping may be made for a specific uniform pipe friction loss knowing the quality and characteristics of the water to be conveyed and the effects that the water will have upon the internal pipe surface in extended service.

Chart 3-1(a)
GALVANIZED IRON AND STEEL STANDARD WEIGHT PIPE (ASTM A72, A120)
Fairly Rough Surface Condition

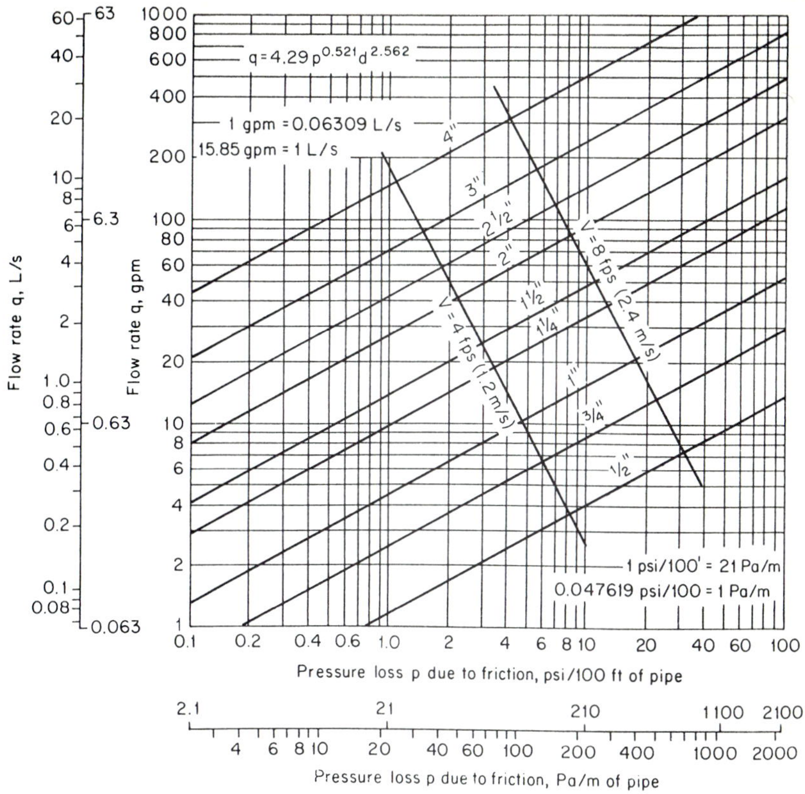

As a guide in selecting the appropriate chart to apply with regard to the quality and characteristics of the water to be conveyed, recommendations are as follows:

FOR RED BRASS AND COPPER PIPE, AND COPPER WATER TUBE

Where the water supply is relatively noncorrosive to the pipe material and is not scale-forming, apply the charts applicable to the "fairly smooth" surface condition.

Where the water supply is moderately scale-forming, apply the charts applicable to the "fairly rough" surface condition.

Where the water has an appreciable scale-forming tendency and would otherwise result in clogging of small-diameter pipes, and in reducing the effective diameter of other piping in the system,

Chart 3-1(b)

GALVANIZED IRON AND STEEL STANDARD WEIGHT PIPE (ASTM A72, A120)

Rough Surface Condition

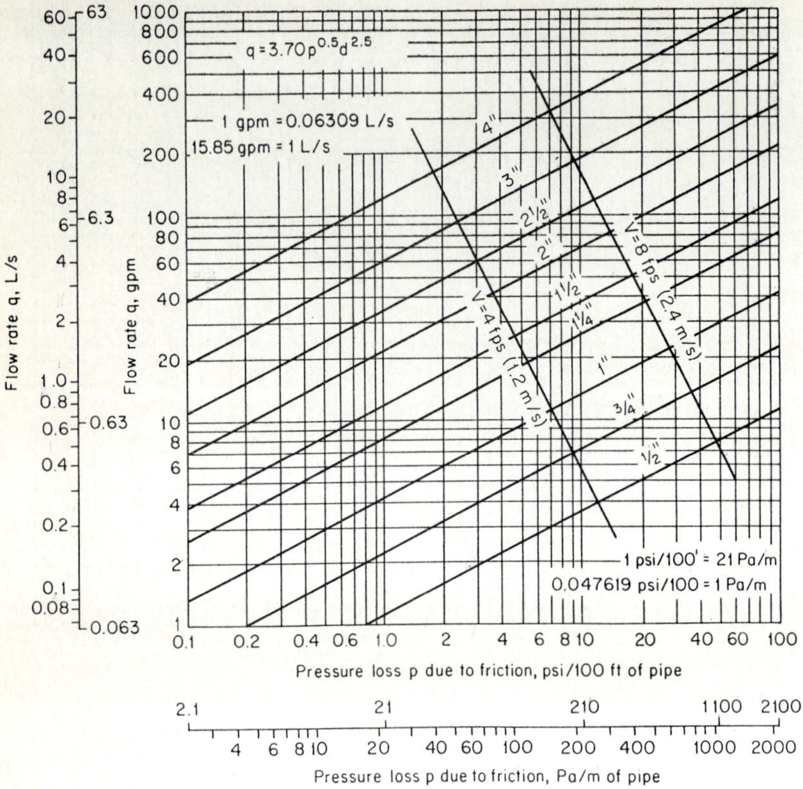

allow one standard pipe size larger than would be determined applying the "fairly smooth" chart.

FOR GALVANIZED IRON AND STEEL PIPE

Where the water supply is relatively noncorrosive to the pipe material and is not scale-forming, apply the chart applicable to the "fairly rough" surface condition.

Where the water supply is moderately corrosive or moderately scale-forming, apply the chart applicable to the "rough" surface condition.

Where the water supply is significantly corrosive to iron or steel or has an appreciable scale-forming tendency and would otherwise

Chart 3-2(a)
COPPER AND BRASS PIPE, S.P.S. (ASTM B42, B43)
Fairly Smooth Surface Condition

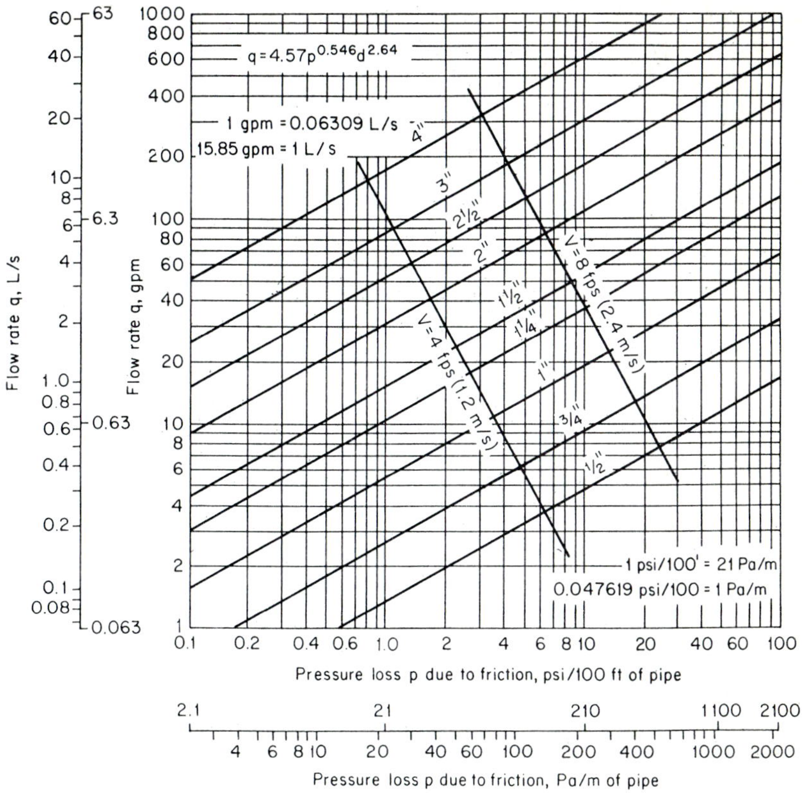

result in clogging of small-diameter pipes, and in reducing the effective diameter of other piping in the system, allow one standard pipe size larger than would be determined applying the "fairly rough" chart.

FOR PLASTIC PIPE (PE, ABS, AND PVC)

Where the water supply is relatively not scale-forming, apply the chart applicable to the "fairly smooth" surface condition.

Where the water supply is moderately scale-forming, apply the chart applicable to the "fairly rough" surface condition.

Where the water has an appreciable scale-forming tendency and would otherwise result in clogging of small-diameter pipes, and

Chart 3-2(b)
COPPER AND BRASS PIPE, S.P.S. (ASTM B42, B43)
Fairly Rough Surface Condition

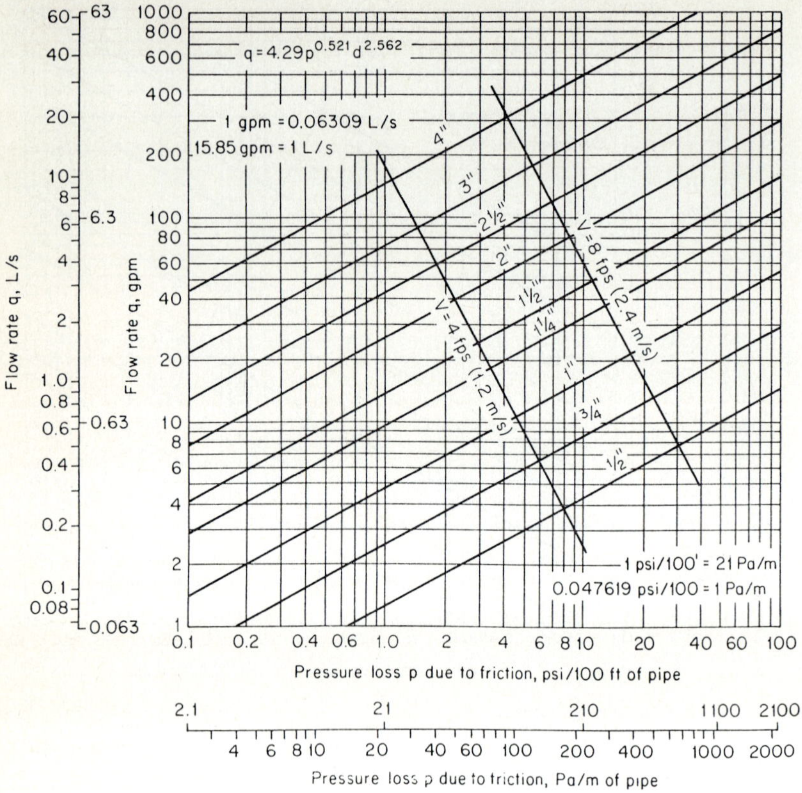

in reducing the effective diameter of other piping in the system, allow one standard pipe size larger than would be determined applying the "fairly smooth" chart.

LIMITATION OF VELOCITY

Consideration of Velocity in Design Velocity of flow through water supply piping during periods of peak demand is an important factor to consider in the design of building water supply systems. Limitation of water velocity should be observed in order to avoid objectionable noise effects in systems, shock damage to piping, equipment, tanks, coils and joints, and accelerated deterioration and eventual failure of piping from erosion-corrosion.

Chart 3-3(a)
THREADLESS RED BRASS AND COPPER PIPE (ASTM B302)
Fairly Smooth Surface Condition

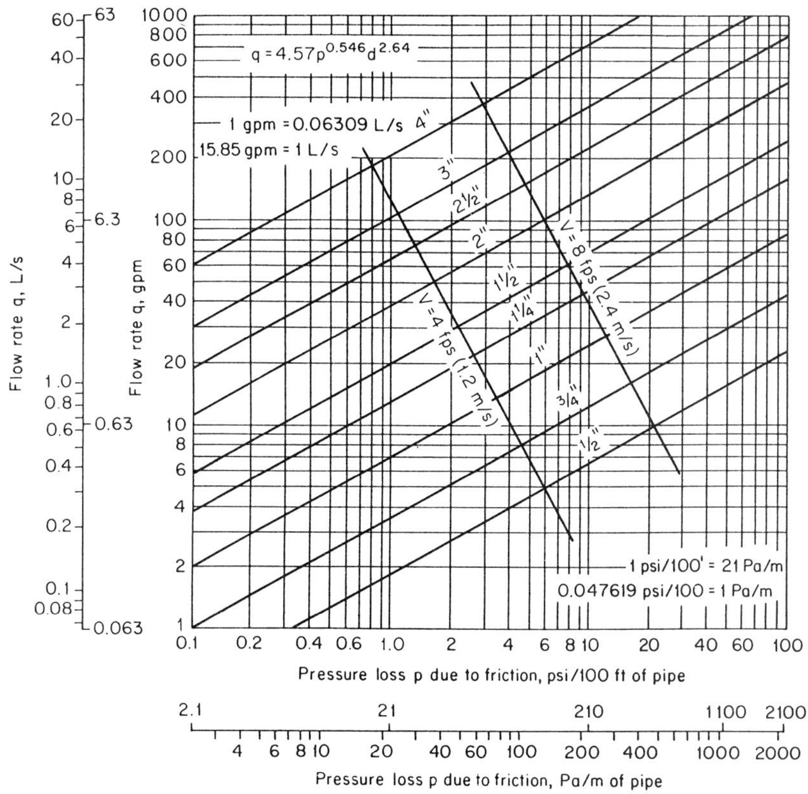

Good Engineering Practice In accordance with good engineering practice, it is recommended that maximum velocity in water supply piping be limited to no more than 8 fps (2.4 m/s). This is deemed essential in order to avoid such objectionable effects as the production of whistling line noise, the occurrence of cavitation, and associated excessive noise in fittings and valves.

It is also recommended that maximum velocity be limited to no more than 4 fps (1.2 m/s) in branch piping from mains, headers, and risers to water outlets at which the supply is controlled by means of quick-closing devices such as an automatic flush valve, solenoid valve, pneumatic valve, or a quick-closing valve or faucet of the self-closing, push-pull, push-button, or other similar type. This limitation is deemed necessary in order to avoid development of excessive and damaging

Chart 3-3(b)
THREADLESS RED BRASS AND COPPER PIPE (ASTM B302)
Fairly Rough Surface Condition

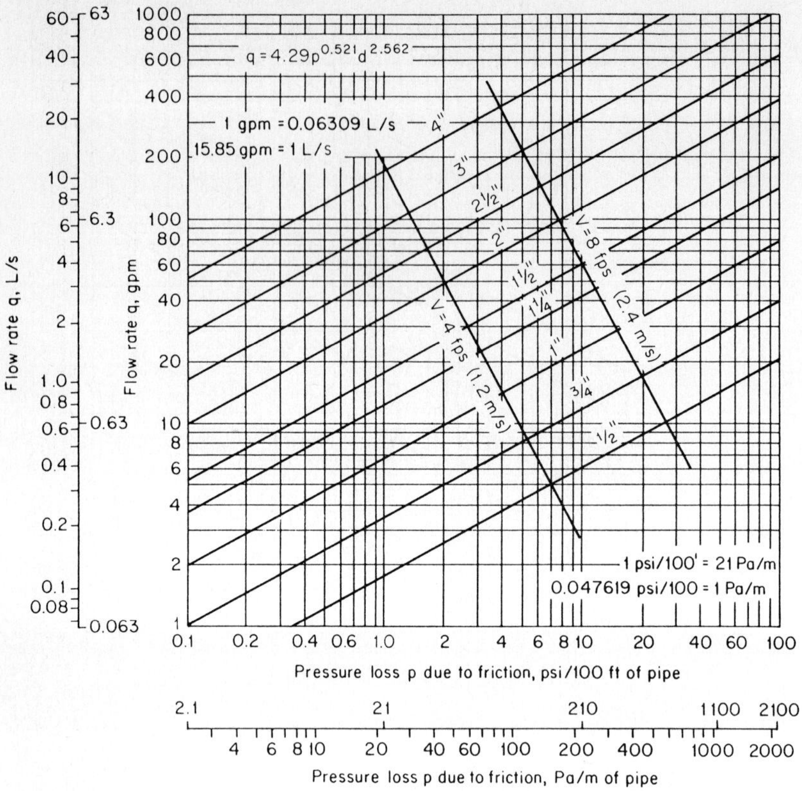

shock pressures in the piping and equipment when flow is suddenly shut off. Many items of plumbing equipment and systems are not designed to withstand the very high shock pressures which may occur as the result of sudden cessation of high-velocity flow in piping.

Manufacturers' Recommendations for Avoiding Erosion-Corrosion

Velocity limits recommended by pipe manufacturers to avoid accelerated deterioration of their piping materials due to erosion-corrosion should be observed. Recent research studies have shown that extreme turbulence accompanying high flow velocities is an important factor in causing erosion-corrosion, and that this is especially prone to occur where the water supply has a high carbon dioxide content (i.e., in excess of 10 ppm), and where it has been softened to zero hardness. Another important factor is very high water temperature (i.e., in excess of 150°F (66°C).

Chart 3-4(a)
COPPER WATER TUBE, TYPE K (ASTM B88)
Fairly Smooth Surface Condition

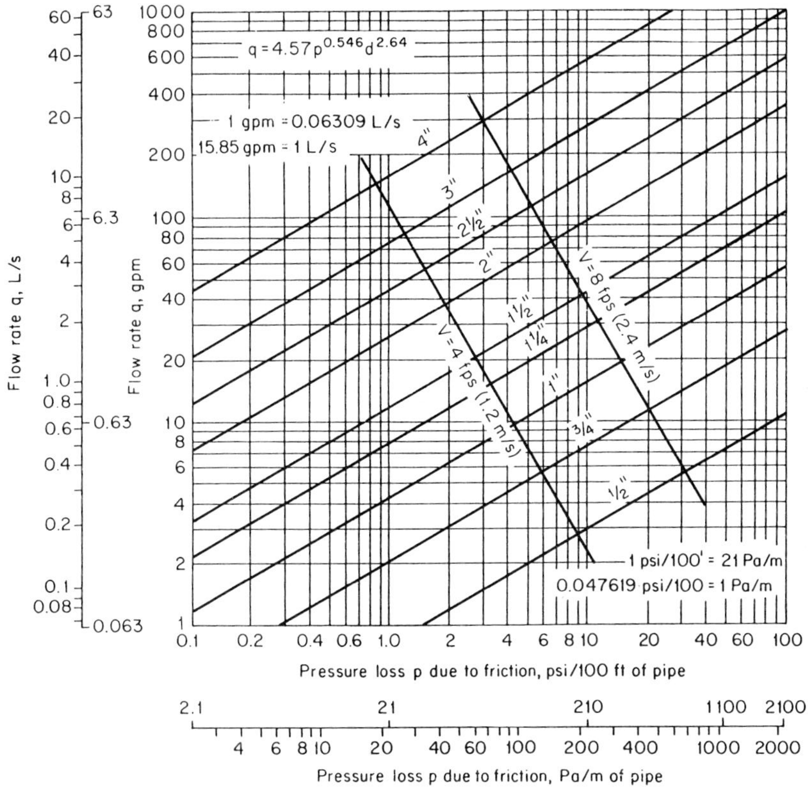

To control erosion-corrosion effects in copper water tube, and copper and brass pipe, pipe manufacturers' recommendations are as follows:

1. Where the water supply has a pH value higher than 6.9 and a positive scale-forming tendency, such as may be shown by a positive Langelier index, velocity should be limited to no more than 8 fps (2.4 m/s).
2. Where the water supply has a pH value lower than 6.9 and may be classified as aggressively corrosive or where the water supply has been softened to zero hardness by passage through a water softener, velocity should be limited to no more than 4 fps (1.2 m/s).
3. The 4 fps (1.2 m/s) velocity limit should be applied to all hot

Chart 3-4(b)
COPPER WATER TUBE, TYPE K (ASTM B88)
Fairly Rough Surface Condition

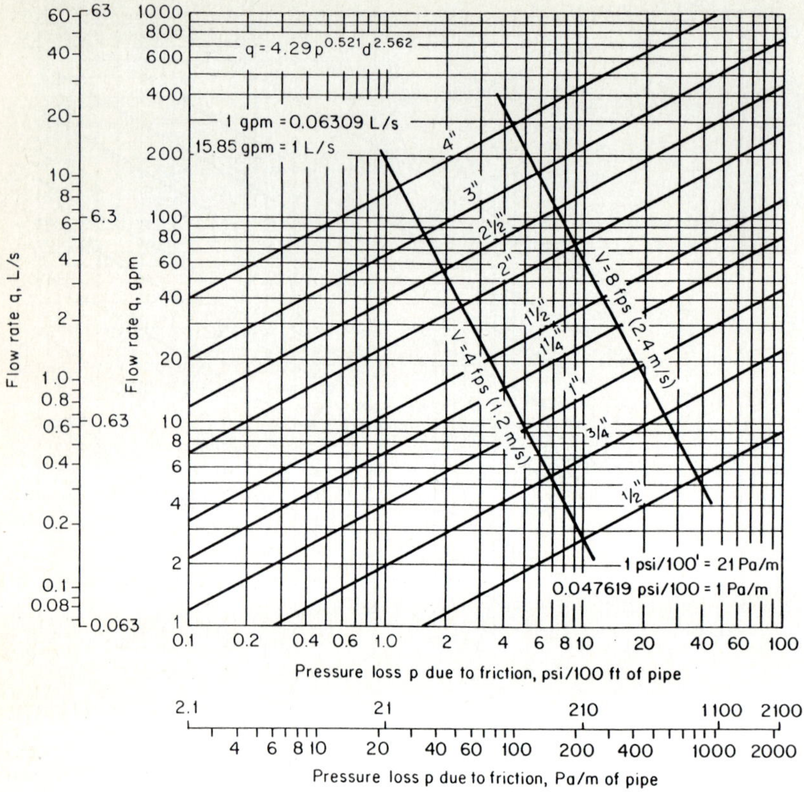

water supply piping conveying water at a temperature above 150°F (65.6°C) because of the accelerated rate of corrosion at such higher temperature.

Recommendations for Minimizing Cost of Pumping Velocity limitation is generally advisable and recommended in the sizing of inlet and outlet piping for water supply pumps. Frictional losses in such piping affect the cost of pumping, and consequently should be reduced to a reasonable minimum. The general recommendation in this instance is to limit velocity in both inlet and outlet piping for water supply pumps to no more than 4 fps (1.2 m/s). This may also be applied advantageously in the case of constant-pressure booster pump water supply systems.

Chart 3-5(a)
COPPER WATER TUBE, TYPE L (ASTM B88)
Fairly Smooth Surface Condition

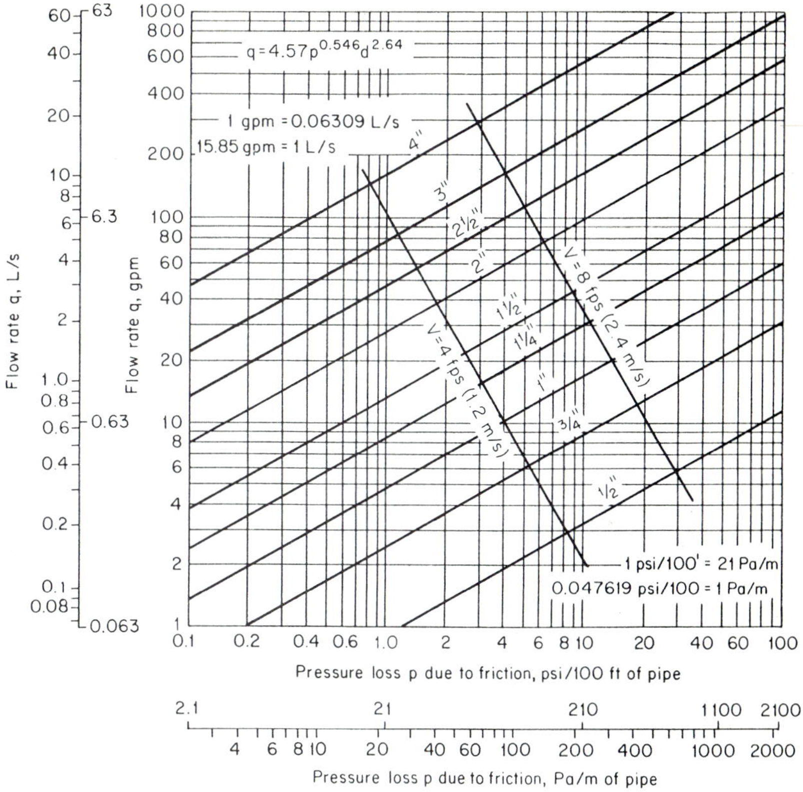

SIZING TABLES BASED ON VELOCITY LIMITATION

To provide a simplified guide for easy application by plumbing designers, and for plumbing plan examiners and inspectors to use in checking pipe sizes for building water supply systems, special sizing tables have been developed. They are presented here in Tables 3-3 through 3-8.

The tables show, for velocity limits of 4 fps (1.2 m/s) and 8 fps (2.4 m/s), the number of water supply fixture units of load permissible for each size and kind of pipe in common use. Hence, the tables may be used in much the same manner as tables currently applied for sizing drainage and vent systems.

An explanation of these tables and how they were developed may serve as an aid to users. There are two tables for each kind of pipe,

Chart 3-5(b)
COPPER WATER TUBE, TYPE L (ASTM B88)
Fairly Rough Surface Condition

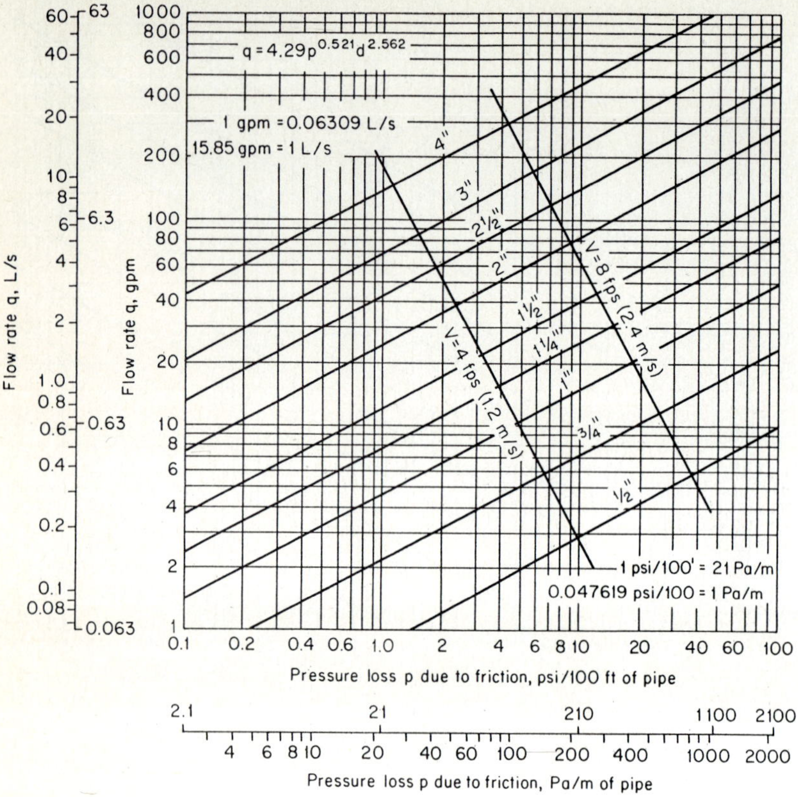

customary U.S. units and SI units. As an example, in Table 3-4*a* the first column lists the nominal pipe sizes, while the second column shows the actual internal pipe diameter as stated in the applicable ASTM standard for the pipe. Under the heading "Velocity = 4 fps" the column headed "Flow" shows the rate of flow, in gallons per minute, calculated for a velocity of 4 fps for each size of pipe based on the actual internal diameter. The column headed "Friction" shows the amount of pipe friction calculated for that rate of flow applying the formula noted beneath the table. Between these two columns are columns A and B, which show load values in terms of water supply fixture units. Column A applies to piping which does not supply flush valves (Flushometers). The load values shown in this column were determined as corresponding to the calculated flow rates using Table 3-2, selecting values as for supply

Chart 3-6(a)

SCHEDULE 40 PLASTIC PIPE: PE (ASTM D2104), ABS (ASTM D1527), PVC (ASTM D1785)

Fairly Smooth Surface Condition

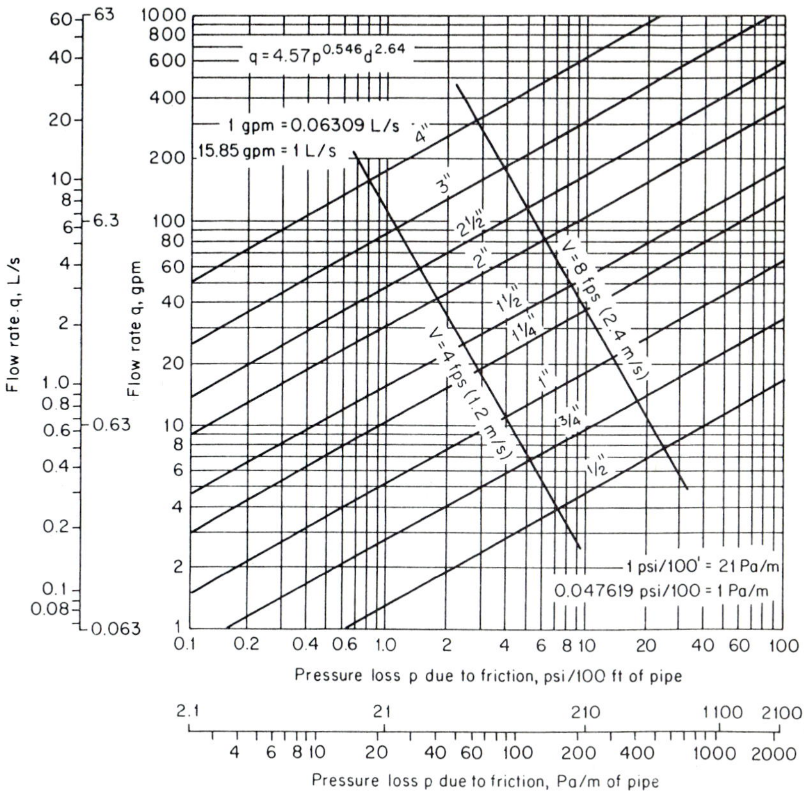

systems predominantly for flush tanks. Column B applies to piping which supplies flush valves. The load values shown therein were similarly determined as corresponding to the same calculated flow rates, using the same table, but selecting such values as for supply systems predominantly for flush valves.

Values shown in the tables for the quantity rate of flow were calculated using the adapted flow formula, which is as follows:

$$q = 2.448\, d^2 V$$

where q = quantity rate of flow, gpm

d = actual inside diameter of pipe, in

V = velocity of flow, fps

Chart 3-6(b)
SCHEDULE 40 PLASTIC PIPE: PE (ASTM D2104), ABS (ASTM D1527), PVC (ASTM D1785)
Fairly Rough Surface Condition

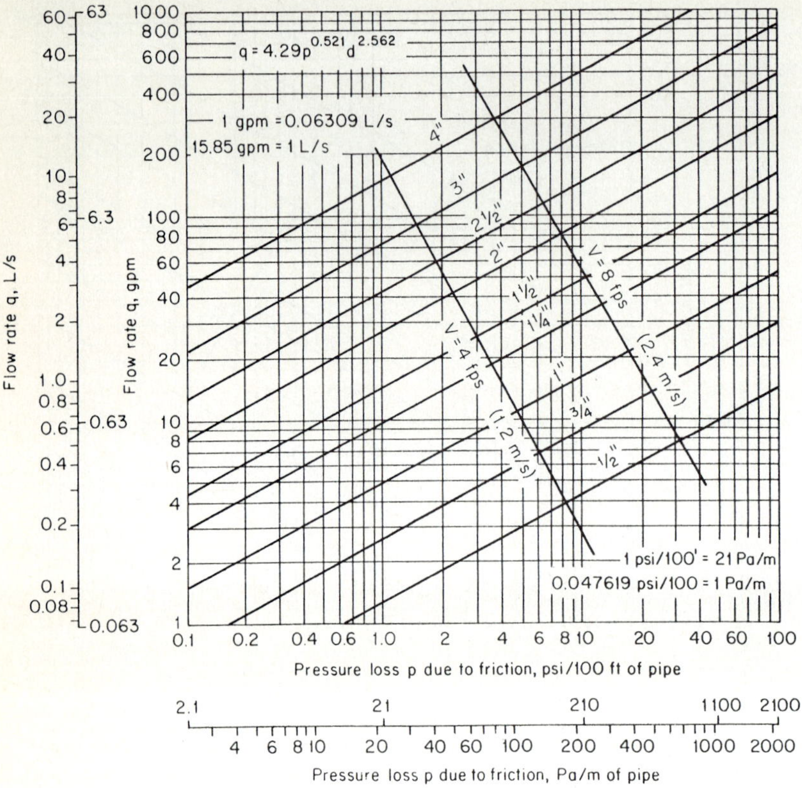

SIMPLIFIED METHOD FOR SIZING SYSTEMS

Application A simplified method for sizing building water supply systems in accordance with demand load, in terms of water supply fixture units, has been found in almost every case to constitute a complete and proper method for adequately sizing the water supply systems of a specific category of buildings. This category includes at least 95 percent of all residential buildings erected per year, and a significant percentage of buildings of other types of occupancy.

In this category are all buildings supplied from a source at which the minimum available water pressure is more than adequate for supplying the highest and most remote fixtures satisfactorily during peak demand, provided the water supply system is sized in accordance with

Table 3-3a
SIZING TABLE BASED ON VELOCITY LIMITATION—U.S. CUSTOMARY UNITS
Galvanized Iron and Steel Pipe, Standard Pipe Size

		Velocity = 4 fps				Velocity = 8 fps			
Nominal size, in	Actual ID, in	Flow q, gpm	Load WSFU (col. A)*	Load WSFU (col. B)†	Friction p, psi/100 ft‡	Flow q, gpm	Load WSFU (col. A)*	Load WSFU (col. B)†	Friction p, psi/100 ft‡
½	0.622	3.8	1.5		8.2	7.6	3.7		31.0
¾	0.824	6.7	3.0		6.0	13.4	8.4		22.5
1	1.049	10.8	6.1		4.6	21.6	25.3	7.7	17.2
1¼	1.380	18.6	17.5	6.0	3.4	37.2	77.3	23.7	12.8
1½	1.610	25.4	37.0	9.3	2.9	50.8	132.3	52.0	10.8
2	2.067	41.8	93.0	29.8	2.2	83.6	293.0	171.6	8.4
2½	2.469	59.8	174.0	75.6	1.8	119.6	477.0	361.0	6.8
3	3.068	92.0	335.0	209.0	1.4	184.0	842.0	806.0	5.4
4	4.026	158.6	688.0	615.0	1.1	317.0	1930.0	1930.0	4.1

* Col. A applies to piping which does not supply flush valves.

† Col. B applies to piping which supplies flush valves.

‡ Friction loss p, corresponding to flow rate q, for piping having fairly smooth surface condition after extended service applying the formula

$$q = 4.57\, p^{0.546}\, d^{2.64}$$

Table 3-3*b*

SIZING TABLE BASED ON VELOCITY LIMITATION—SI UNITS

Galvanized Iron and Steel Pipe, Standard Pipe Size

		Velocity = 1.2 m/s				Velocity = 2.4 m/s			
Nominal size, mm	Actual ID, mm	Flow q, L/s	Load WSFU (col. A)*	Load WSFU (col. B)†	Friction p, Pa/m‡	Flow q, L/s	Load WSFU (col. A)*	Load WSFU (col. B)†	Friction p, Pa/m‡
12.7	15.8	0.23	1.5		172.3	0.47	3.7		651.5
19.0	20.9	0.42	3.0		126.1	0.84	8.4		472.8
25.4	26.6	0.68	6.1		96.7	1.36	25.3	7.7	361.5
31.8	35.1	1.17	17.5	6.0	71.5	2.34	77.3	23.7	269.0
38.1	40.9	1.60	37.0	9.3	60.9	3.20	132.3	52.0	227.0
50.8	52.5	2.63	93.0	29.8	46.2	5.27	293.0	171.6	176.5
63.5	62.7	3.77	174.0	75.6	37.8	7.54	477.0	361.0	142.9
76.2	77.7	5.80	335.0	209.0	29.4	11.60	842.0	806.0	113.5
102.0	102.3	10.00	688.0	615.0	23.1	20.01	1930.0	1930.0	86.2

* Col. A applies to piping which does not supply flush valves.

† Col. B applies to piping which supplies flush valves.

‡ Friction loss p, corresponding to flow rate q, for piping having fairly smooth surface condition after extended service applying the formula

$$q = 4.57\, p^{0.546}\, d^{2.64}$$

Table 3-4*a*
SIZING TABLE BASED ON VELOCITY LIMITATION—U.S. CUSTOMARY UNITS
Copper and Brass Pipe, Standard Pipe Size

		Velocity = 4 fps				Velocity = 8 fps			
Nominal size, in	Actual ID, in	Flow q, gpm	Load WSFU (col. A)*	Load WSFU (col. B)†	Friction p, psi/100 ft‡	Flow q, gpm	Load WSFU (col. A)*	Load WSFU (col. B)†	Friction p, psi/100 ft‡
½	0.625	3.8	1.5		6.8	7.0	3.7		24.2
¾	0.822	6.6	3.0		5.1	3.2	8.4		18.0
1	1.062	11.0	6.3		3.7	22.0	26.4	8.0	13.3
1¼	1.368	18.3	16.8	6.4	2.8	36.6	75.0	22.7	10.0
1½	1.600	25.2	36.3	9.3	2.3	50.4	130.0	51.0	8.4
2	2.062	41.6	92.0	29.5	1.7	83.2	291.0	170.0	6.2
2½	2.500	61.2	181.0	80.0	1.4	122.4	492.0	376.0	4.9
3	3.062	92.0	335.0	209.0	1.1	184.0	842.0	807.0	3.9
4	4.000	158.0	685.0	611.0	0.8	316.0	1920.0	1920.0	2.9

* Col. A applies to piping which does not supply flush valves.

† Col. B applies to piping which supplies flush valves.

‡ Friction loss p corresponding to flow rate q for piping having fairly smooth surface condition after extended service applying the formula

$$q = 4.57\, p^{0.546}\, d^{2.64}$$

Table 3-4*b*

SIZING TABLE BASED ON VELOCITY LIMITATION—SI UNITS

Copper and Brass Pipe, Standard Pipe Size

		Velocity = 1.2 m/s				Velocity = 2.4 m/s			
Nominal size, mm	Actual ID, mm	Flow q, L/s	Load WSFU (col. A)*	Load WSFU (col. B)†	Friction p, Pa/m‡	Flow q, L/s	Load WSFU (col. A)*	Load WSFU (col. B)†	Friction p, Pa/m‡
12.7	15.9	0.24	1.5		143.0	0.48	3.7		508.6
19.0	20.9	0.42	3.0		107.2	0.83	8.4		378.3
25.4	27.0	0.69	6.3		71.8	1.39	26.4	8.0	279.5
31.8	34.7	1.15	16.8	6.4	58.8	2.31	75.0	22.7	210.2
38.1	40.6	1.59	36.3	9.3	48.3	3.18	130.0	51.0	176.5
50.8	52.4	2.62	92.0	29.0	35.7	5.25	291.0	170.0	130.3
63.5	63.5	3.86	181.0	80.0	29.4	7.72	492.0	376.0	103.0
76.2	77.8	5.80	335.0	209.0	23.1	11.61	842.0	807.0	82.0
102.0	101.6	9.97	685.0	611.0	16.8	19.94	1920.0	1920.0	61.0

* Col. A applies to piping which does not supply flush valves.

† Col. B applies to piping which supplies flush valves.

‡ Friction loss p corresponding to flow rate q for piping having fairly smooth surface condition after extended service applying the formula

$$q = 4.57\ p^{0.546}\ d^{2.64}$$

Table 3-5*a*

SIZING TABLE BASED ON VELOCITY LIMITATION—U.S. CUSTOMARY UNITS

Threadless Copper and Red Brass Pipe (TP)

		Velocity = 4 fps				Velocity = 8 fps			
Nominal size, in	Actual ID, in	Flow q, gpm	Load WSFU (col. A)*	Load WSFU (col. B)†	Friction p, psi/100 ft‡	Flow q, gpm	Load WSFU (col. A)*	Load WSFU (col. B)†	Friction p, psi/100 ft‡
½	0.710	4.9	2.0		5.9	9.8	5.3		20.8
¾	0.920	8.3	4.2		4.4	16.6	13.2	5.7	15.5
1	1.185	13.7	9.0		3.3	27.4	44.0	10.5	11.7
1¼	1.530	22.9	28.9	8.3	2.4	45.8	110.0	40.0	8.5
1½	1.770	30.6	55.0	14.5	2.1	61.2	181.0	80.0	7.2
2	2.245	49.4	126.0	48.5	1.6	98.8	369.0	240.0	5.6
2½	2.745	74.0	245.0	125.0	1.3	148.0	631.0	537.0	4.4
3	3.334	109.0	421.0	305.0	1.0	218.0	1081.0	1081.0	3.5
4	4.286	180.0	816.0	774.0	0.8	360.0	2318.0	2318.0	2.6

* Col. A applies to piping which does not supply flush valves.

† Col. B applies to piping which supplies flush valves.

‡ Friction loss p corresponding to flow rate q for piping having fairly smooth surface condition after extended service applying the formula

$$q = 4.57\, p^{0.546}\, d^{2.64}$$

Table 3-5*b*
SIZING TABLE BASED ON VELOCITY LIMITATION—SI UNITS
Threadless Copper and Red Brass Pipe (TP)

		Velocity = 1.2 m/s				Velocity = 2.4 m/s			
Nominal size, mm	Actual ID, mm	Flow q, L/s	Load WSFU (col. A)*	Load WSFU (col. B)†	Friction p, Pa/m‡	Flow q, L/s	Load WSFU (col. A)*	Load WSFU (col. B)†	Friction p, Pa/m‡
12.7	18.0	0.31	2.0		124.0	0.62	5.3		437.1
19.0	23.4	0.52	4.2		92.5	1.05	13.2	5.7	325.7
25.4	30.1	0.86	9.0		69.4	1.73	44.0	10.5	245.9
31.8	38.9	1.44	28.9	8.3	50.4	2.89	110.0	40.0	178.6
38.1	47.0	1.91	55.0	14.5	44.1	3.86	181.0	80.0	151.3
50.8	57.0	3.11	126.0	48.5	33.6	6.23	369.0	240.0	117.7
63.5	69.7	4.6	245.0	125.0	27.3	9.34	631.0	537.0	92.5
76.2	84.7	6.8	421.0	305.0	21.0	13.75	1081.0	1081.0	73.6
102.0	108.9	11.4	816.0	774.0	16.8	22.71	2318.0	2318.0	54.6

* Col. A applies to piping which does not supply flush valves.

† Col. B applies to piping which supplies flush valves.

‡ Friction loss p corresponding to flow rate q for piping having fairly smooth surface condition after extended service applying the formula

$$q = 4.57\, p^{0.546}\, d^{2.64}$$

Table 3-6*a*

SIZING TABLE BASED ON VELOCITY LIMITATION—U.S. CUSTOMARY UNITS

Copper Water Tube, Type K

		Velocity = 4 fps				Velocity = 8 fps			
Nominal size, in	Actual ID, in	Flow q, gpm	Load WSFU (col. A)*	Load WSFU (col. B)†	Friction p, psi/100 ft‡	Flow q, gpm	Load WSFU (col. A)*	Load WSFU (col. b)†	Friction p, psi/100 ft‡
½	0.527	2.7	0.75		8.5	5.4	2.3		31.0
¾	0.745	5.5	2.3		5.6	11.0	6.3		20.2
1	0.995	9.7	5.3		4.1	19.4	19.5	5.8	14.4
1¼	1.245	15.2	10.8	5.0	3.1	30.4	54.0	14.0	11.1
1½	1.481	21.5	25.0	7.8	2.6	43.0	98.0	34.0	9.2
2	1.959	37.6	78.0	24.0	1.8	75.2	251.0	130.0	6.5
2½	2.435	58.2	166.0	69.0	1.4	116.4	460.0	340.0	5.2
3	2.907	82.8	289.0	161.0	1.2	165.6	725.0	663.0	4.2
4	3.857	146.0	609.0	528.0	0.8	292.0	1705.0	1705.0	3.0

* Col. A applies to piping which does not supply flush valves.

† Col. B applies to piping which supplies flush valves.

‡ Friction loss p corresponding to flow rate q for piping having fairly smooth surface condition after extended service applying the formula

$$q = 4.57\, p^{0.546}\, d^{2.64}$$

Table 3-6*b*
SIZING TABLE BASED ON VELOCITY LIMITATION—SI UNITS
Copper Water Tube, Type K

		Velocity = 1.2 m/s				Velocity = 2.4 m/s			
Nominal size, mm	Actual ID, mm	Flow q, L/s	Load WSFU (col. A)*	Load WSFU (col. B)†	Friction p, Pa/m‡	Flow q, L/s	Load WSFU (col. A)*	Load WSFU (col. B)†	Friction p, Pa/m‡
12.7	13.4	0.17	0.75		178.6	0.34	2.3		651.5
19.0	18.9	0.34	2.3		117.7	0.69	6.3		424.5
25.4	25.3	0.61	5.3		86.2	1.22	19.5	5.8	302.6
31.8	31.6	0.95	10.8	5.0	65.1	1.91	54.0	14.0	233.3
38.1	37.6	1.35	25.0	7.8	54.6	2.71	98.0	34.0	193.3
50.8	49.8	2.37	78.0	24.0	37.8	4.74	251.0	130.0	136.6
63.3	61.8	3.67	166.0	69.0	29.4	7.34	460.0	340.0	109.3
76.2	73.8	5.22	289.0	161.0	25.2	10.44	725.0	663.0	88.3
102.0	98.0	9.21	609.0	528.0	16.8	18.42	1705.0	1705.0	63.0

* Col. A applies to piping which does not supply flush valves.

† Col. B applies to piping which supplies flush valves.

‡ Friction loss p corresponding to flow rate q for piping having fairly smooth surface condition after extended service applying the formula

$$q = 4.57\, p^{0.546}\, d^{2.64}$$

Table 3-7*a*
SIZING TABLE BASED ON VELOCITY LIMITATION—U.S. CUSTOMARY UNITS
Copper Water Tube, Type L

		Velocity = 4 fps				Velocity = 8 fps			
Nominal size, in	Actual ID, in	Flow q, gpm	Load WSFU (col. A)*	Load WSFU (col. B)†	Friction p, psi/100 ft‡	Flow q, gpm	Load WSFU (col. A)*	Load WSFU (col. B)†	Friction p, psi/100 ft‡
½	0.545	2.9	1.0		8.2	5.8	2.5		29.0
¾	0.785	6.0	2.5		5.2	12.0	7.3		18.7
1	1.025	10.3	5.5		3.9	20.6	22.5	7.0	13.7
1¼	1.265	15.7	11.5	5.0	3.0	31.4	58.0	15.5	10.7
1½	1.505	22.8	28.5	8.0	2.5	45.6	109.0	38.0	8.7
2	1.985	38.6	82.0	26.0	1.8	77.2	261.0	138.0	6.3
2½	2.465	59.5	172.0	75.0	1.4	119.0	474.0	356.0	4.9
3	2.945	85.0	300.0	178.0	1.1	170.0	750.0	692.0	4.0
4	3.905	149.0	636.0	544.0	0.8	298.0	1759.0	1759.0	2.8

* Col. A applies to piping which does not supply flush valves.

† Col. B applies to piping which supplies flush valves.

‡ Friction loss p corresponding to flow rate q for piping having fairly smooth surface condition after extended service applying the formula

$$q = 4.57\, p^{0.546}\, d^{2.64}$$

Table 3-7*b*
SIZING TABLE BASED ON VELOCITY LIMITATION—SI UNITS
Copper Water Tube, Type L

		Velocity = 1.2 m/s				Velocity = 2.4 m/s			
Nominal size, mm	Actual ID, mm	Flow q, L/s	Load WSFU (col. A)*	Load WSFU (col. B)†	Friction p, Pa/m‡	Flow q, L/s	Load WSFU (col. A)*	Load WSFU (col. B)†	Friction p, Pa/m‡
12.7	13.8	0.18	1.0		172.3	0.36	2.5		609.4
19.0	19.9	0.37	2.5		109.3	0.75	7.3		392.9
25.4	26.0	0.64	5.5		80.7	1.29	22.5	7.0	287.9
31.8	31.1	0.99	11.5	5.0	63.0	1.98	58.0	15.5	224.9
38.1	38.2	1.43	28.5	8.0	52.5	2.87	109.0	38.0	182.8
50.8	50.4	2.43	82.0	26.0	37.8	4.87	261.0	138.0	132.4
63.5	62.6	3.75	172.0	75.0	29.4	7.50	474.0	356.0	102.9
76.2	74.8	5.36	300.0	178.0	23.1	10.72	750.0	692.0	84.0
102.0	99.2	9.40	636.0	544.0	16.8	18.80	1759.0	1759.0	58.8

* Col. A applies to piping which does not supply flush valves.

† Col. B applies to piping which supplies flush valves.

‡ Friction loss p corresponding to flow rate q for piping having fairly smooth surface condition after extended service applying the formula

$$q = 4.57\, p^{0.546}\, d^{2.64}$$

Table 3-8*a*

SIZING TABLE BASED ON VELOCITY LIMITATION—U.S. CUSTOMARY UNITS

Schedule 40 Plastic Pipe (PE, PVC, and ABS)

		Velocity = 4 fps				Velocity = 8 fps			
Nominal size, in	Actual ID, in	Flow q, gpm	Load WSFU (col. A)*	Load WSFU (col. B)†	Friction p, psi/100 ft‡	Flow q, gpm	Load WSFU (col. A)*	Load WSFU (col. B)†	Friction p, psi/100 ft‡
½	0.622	3.8	1.5		6.8	7.6	3.7		24.2
¾	0.824	6.7	3.0		5.1	13.4	8.4		18.0
1	1.049	10.8	6.1		3.7	21.6	25.3	7.7	13.2
1¼	1.380	18.6	17.5	6.0	2.8	37.2	77.3	23.7	9.6
1½	1.610	25.4	37.0	9.3	2.3	50.8	132.3	52.0	8.2
2	2.067	41.8	93.0	29.8	1.7	83.6	293.0	171.6	6.1
2½	2.469	59.8	174.0	75.6	1.4	119.6	477.0	361.0	4.8
3	3.068	92.0	335.0	209.0	1.1	184.0	842.0	806.0	3.8
4	4.026	158.6	688.0	615.0	0.8	317.2	1930.0	1930.0	2.8

* Col. A applies to piping which does not supply flush valves.

† Col. B applies to piping which supplies flush valves.

‡ Friction loss p corresponding to flow rate q for piping having fairly smooth surface condition after extended service applying the formula

$$q = 4.57\, p^{0.546}\, d^{2.64}$$

Table 3-8*b*
SIZING TABLE BASED ON VELOCITY LIMITATION—SI UNITS
Schedule 40 Plastic Pipe (PE, PVC, and ABS)

		Velocity = 1.2 m/s				Velocity = 2.4 m/s			
Nominal size, mm	Actual ID, mm	Flow q, L/s	Load WSFU (col. A)*	Load WSFU (col. B)†	Friction p, Pa/m‡	Flow q, L/s	Load WSFU (col. A)*	Load WSFU (col. B)†	Friction p, Pa/m‡
12.7	15.8	0.23	1.5		142.9	0.47	3.7		508.6
19.0	20.9	0.42	3.0		107.2	0.84	8.4		378.3
25.4	26.6	0.68	6.1		77.8	1.36	25.3	7.7	277.4
31.8	35.1	1.17	17.5	6.0	58.8	2.34	77.3	23.7	201.7
38.1	40.9	1.60	37.0	9.3	48.3	3.20	132.3	52.0	172.3
50.8	52.5	2.63	93.0	29.8	35.7	5.27	293.0	171.6	128.2
63.5	62.7	3.77	174.0	75.6	29.4	7.54	477.0	361.0	100.9
76.2	77.7	5.80	335.0	209.0	23.1	11.60	842.0	806.0	79.9
102.0	102.3	10.00	688.0	615.0	16.8	20.01	1930.0	1930.0	58.8

* Col. A applies to piping which does not supply flush valves.

† Col. B applies to piping which supplies flush valves.

‡ Friction loss p corresponding to flow rate q for piping having fairly smooth surface condition after extended service applying the formula

$$q = 4.57\, p^{0.546}\, d^{2.64}$$

the velocity limitations which should be observed. Included are almost all one- and two-family dwellings, most multiple dwellings up to at least three stories in height, and a considerable proportion of commercial and industrial buildings of limited height and area, where supplied from a source at which the minimum available pressure is no less than 40 psi (276 kPa). Under such conditions, the available pressure generally is more than enough for overcoming static head and ordinary pipe friction losses, and pipe friction is not an additional factor to consider in sizing.

Simplified Method Based on Velocity Limitation This method is based solely on the application of velocity limitations that are (1) recognized as good engineering practice; (2) authoritative recommendations issued by manufacturers of piping materials regarding proper use of their products in order to achieve durable performance and avoid failure in service, especially in water areas where the supply is aggressively corrosive; and (3) general recommendations for minimizing the cost of pumping where water supply pumps are provided. These limitations have been detailed in the section "Limitation of Velocity."

STEP-BY-STEP PROCEDURE FOR THE SIMPLIFIED SIZING METHOD

For sizing water supply systems in relatively low buildings, a simplified method has been developed in a step-by-step procedure which may be applied in the design of modern buildings. The procedure consists of seven steps. They are as follows:

1. Obtain all information necessary for establishing a proper basis for sizing the system. Properness of the basis for sizing is contingent upon accuracy and reliability of the information applied. Such information should be obtained from responsible parties and appropriate local authorities recognized as sources of the necessary information. See the preceding discussion under the subject "Preliminary Information."
2. Provide a schematic elevation of the complete water supply system. Show all piping connections in proper sequence and all fixture supplies. Identify all fixtures and risers by means of appropriate letters, numbers, or combinations thereof. Specially identify all piping conveying water at a temperature above 150°F (66°C), and all branch piping to such water outlets as automatic flush valves, solenoid valves, pneumatic valves, or quick-closing valves or faucets. Provide on the schematic elevation all the

necessary information obtained as per step 1, and as discussed under the subject "Preliminary Information."

3. Mark on the schematic elevation, for each section of the complete system, the hot and cold water loads conveyed thereby in terms of water supply fixture units in accordance with Table 3-14.
4. Mark on the schematic elevation, adjacent to all fixture units notations, the demand in gallons per minute or liters per second corresponding to the various fixture unit loads in accordance with Table 3-2.
5. Mark on the schematic elevation, for appropriate sections of the system, the demand in gallons per minute or liters per second for outlets at which demand is deemed continuous, such as outlets for watering gardens, irrigating lawns, air-conditioning apparatus, refrigeration machines, and other similar equipment using water at a relatively continuous rate during peak demand periods. Add the continuous demand to the demand for intermittently used fixtures, and show the total demand at those sections where both types of demand occur.
6. Size all individual fixture supply pipes to water outlets in accordance with the minimum sizes permitted by regulations. Minimum fixture supply pipe sizes for common plumbing fixtures are given in Table 3-15.
7. Size all other parts of the water supply system in accordance with velocity limitations recognized as good engineering practice, with velocity limitations recommended by pipe manufacturers for avoiding accelerated deterioration and failure of their products under various conditions of service, and with velocity limitations generally recommended for minimizing the cost of pumping where water supply pumps are provided. [Sizing tables (Tables 3-3 to 3-8) based on such velocity limitations and showing permissible loads in terms of water supply fixture units for each size and kind of piping material have been provided and may be applied advantageously in this step.]

APPLICATION OF SIMPLIFIED METHOD TO ILLUSTRATIVE PROBLEMS

Problem 1

Draw a schematic elevation, and size the piping of the following water distributing system using the simplified sizing method:

A one-family dwelling, two stories and cellar in height, is to be supplied by direct pressure from an 8-in public water main in which the certified minimum pressure available is 50 psi. Top floor fixture outlets are 20 ft above the public main and require 8 psi flow pressure for satisfactory operation.

Authoritative reports indicate that the public water supply has a pH of 7.2, carbon dioxide content of 7 ppm, and a solids content of 90 ppm. Records show no significant corrosion of copper by the water up to 150°F.

Copper water tube with wrought copper fittings has been selected, and is to be of type K for the water service and of type L for inside the building.

Water supply for the premises is to be metered at the point of entry by a disk-type meter. The system is to be of the upfeed riser type. A 52-gal automatic hot water storage heater, having a rated pressure loss of 1.5 psi at peak demand, is to provide 140°F tank controlled hot water supply.

The most extreme run of piping from the public main to the highest and most remote outlet is 180 ft in developed length, consisting of the following: 100 ft of water service, 45 ft of cold water piping from the water service valve to the hot water storage heater, and 35 ft therefrom to the top floor hot water outlet at the shower. Plans of the entire water supply system are available.

Fixtures provided on the system are as follows:

> Cellar—an automatic laundry washing machine, two hose bib outlets (only one to be used at any one time) at the outside of the building for lawn watering, and one valved outlet for supplying the house heating boiler
>
> First floor—a kitchen equipped with a sink and a domestic dishwasher, and a powder room containing one lavatory and one water closet with flush tank
>
> Second floor—two bathroom groups, one containing a lavatory, bathtub with shower above, and a water closet with flush tank, and a second containing a lavatory, shower stall, and a water closet with flush tank

Solution to Problem 1

Step 1 All information required for establishing a proper design basis has been obtained from appropriate sources.

Step 2 A schematic elevation of the building water supply system is shown in Fig. 3-1. This drawing was developed using the plans of the

DESIGN BASIS

PIPING:
Copper water tube with wrought copper fittings, type K for water service, type L for inside building.

PUBLIC SUPPLY SYSTEM:
8" public main located in street in in front of building. Certified minimum available pressure is: 50 psi.

WATER CHARACTERISTICS:
Records show no significant corrosion of copper by the water up to 150 °F.

WATER ANALYSIS:

pH	7.2
CO_2	7 ppm
Dissolved solids	90 ppm

ELEVATIONS:

Curb (as datum)	0.00'
Public main	-4.00'
Cellar floor	-7.00'
First floor	+2.00'
Second floor	+12.00'
Highest outlet	+15.00'

DEVELOPED LENGTHS OF PIPING FROM PUBLIC MAIN TO HIGHEST AND FARTHEST OUTLET:
Sections: PM-A = 100'; A-B = 40'; B-C = 5'; C-D = 5'; D-E = 10'; E-F = 10'; F-G = 10'; TOTAL DL = 180'.

HOT WATER TEMPERATURE:
140°F system, tank control.

DESIGN VELOCITY LIMITS:
8 fps for all piping, except 4 fps for branches to quick-closing valves (noted by *).

KEY FOR NOTATIONS MADE IN SIZING PROCEDURE
Load in water supply fixture units is shown unenclosed.
Load in gpm of flow is shown in ( ).
Continuous load in gpm of flow is shown in (_).
Sizes selected are shown in ["].

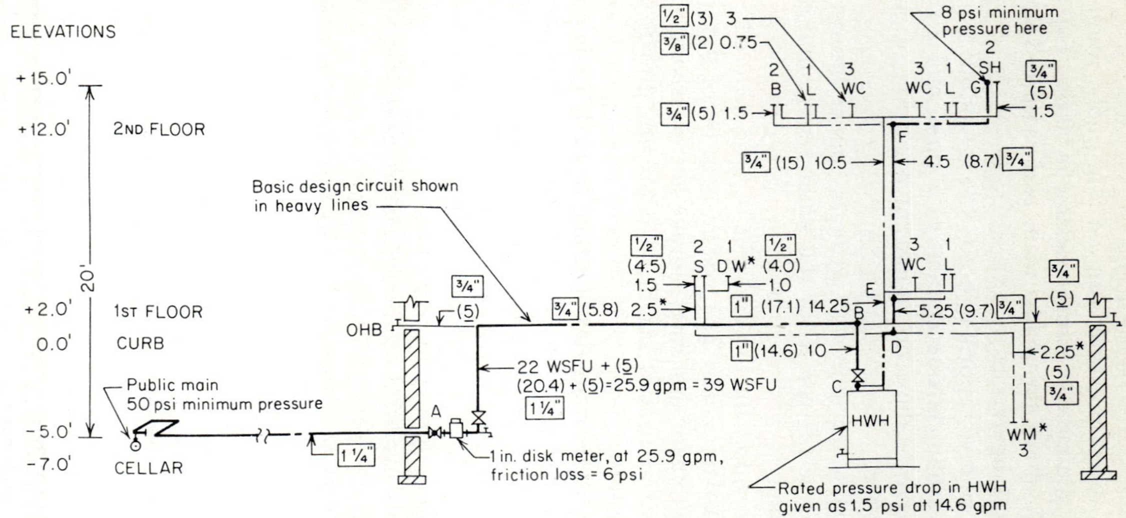

Fig. 3-1 Water supply system design sheet for Prob. 1.

system. All piping connections have been shown in proper sequence as per the plans, and the developed lengths of each section of the basic design circuit determined therefrom. All information required for establishing a proper design basis has been shown on the left side of the design sheet. Fixtures have been identified by letters, and those fixtures and branches having quick-closing valve outlets have been specially identified by means of an asterisk.

Step 3 For each section of the system, notations have been made showing the hot and cold water loads conveyed thereby in terms of water supply fixture units. Fixture unit values have been shown unenclosed by parentheses.

Step 4 Adjacent to fixture unit load notations, the demand in gallons per minute corresponding to such loads has been shown in parentheses. The demand in gpm was determined from Table 3-2, applying the values shown therein under the heading, "Supply Systems Predominantly for Flush Tanks," in view of the fact that no flush valves (Flushometers) are involved in this water supply system.

Step 5 The continuous demand posed by the two outside hose bibs, only one of which is to be used at a time, has been shown on the design sheet. This demand has been specially underlined and included in parentheses as gallons per minute where added to fixture units of load. The normal demand posed by a hose bib was obtained from Table 3-1.

Step 6 All individual fixture supply pipes to water outlets have been sized on the design sheet in accordance with the minimum sizes shown under the subject, Standard Code Regulations, in Table 3-15.

Step 7 All other parts of the system have been sized in accordance with the velocity limitations established for this system as the proper basis for design, i.e., 8 fps for all piping, except 4 fps for branches to quick-closing valves as noted by asterisks on the design sheet. Sizing was done in accordance with total fixture units of load corresponding to total demand in each section. For those sections of the cold water header in the cellar which convey both demand of intermittently used fixtures and continuous demand of a hose bib, the total demand in gallons per minute was converted to equivalent water supply fixture units of load and proper sizes were determined therefor, although proper sizing could also have been done simply on the basis of demand rates in gallons per minute. Sizing was done using Tables 3-3 through 3-8, and specifically those tables dealing with copper water tube, type K, for sizing the water service pipe, and with copper water tube, type L, for sizing piping inside the building.

SUPPLEMENTARY CHECK RE PIPE FRICTION LOSSES IN BDC

To prove the adequacy of the simplified sizing method as applied in this problem, a supplementary check has been made regarding the total pipe friction losses in the main lines and risers for the longest run of piping from the public water main to the highest and most remote water outlet. This run of piping was shown in heavy lines on the design sheet and letters marked thereon to denote the various sections of the run, the BDC.

The certified minimum pressure available at the public main was given as 50 psi. After appropriate deductions therefrom, the maximum pressure available for pipe friction losses during peak demand in the BDC was determined to be 25.8 psi. The appropriate deductions were as follows: 8.7 psi for static head loss due to rise from the public main to the highest outlet, 8.0 psi minimum pressure required at the top outlet, 6.0 psi friction loss through the 1-in disk meter at 25.9 gpm flow rate, and 1.5 psi friction loss rated for 14.6 gpm flow rate through the 52-gal hot water storage heater.

Table 3-9 shows the losses determined for each section of the circuit. Total pipe friction loss for the BDC was determined to be 18.6 psi, a total of 7.2 psi less than the 25.8 psi previously determined as the maximum pressure available for pipe friction losses during peak demand in the BDC. This proves that the sizes determined in this problem on the basis of velocity limitations exclusively are adequate.

Problem 2

Draw a schematic elevation, and size the piping of the following water distributing system using the simplified sizing method:

A nine-family multiple dwelling, three stories and cellar in height, fronts on a public street and is to be supplied by direct street pressure from a 10-in public water main in which the maximum pressure is 55 psi (379.5 kPa) and the minimum pressure available is 40 psi (276 kPa) as certified by a hydrant flow test. Top floor fixture outlets are 29 ft (87 m) above the public main and require 8 psi (55.2 kPa) flow pressure for satisfactory operation.

Water analysis of the public supply is as follows:

pH 7.1	Hardness 23 ppm
Iron 0.07 ppm	Alkalinity 13 ppm

Table 3-9
PIPE FRICTION LOSSES IN BASIC DESIGN CIRCUIT FOR PROB. 1
Copper Water Tube

Piping section	Tube type	Tube size, in	Developed length, ft	Equivalent length of fittings and valves, ft	Total equivalent length of section, ft	Flow rate *q* gpm	Friction (Charts 6-4*a*, 6-5*a*), psi/100′	Friction loss, psi
PM–A	K	1¼	100	1, 1¼″ tap at 1.9′ 1, 1¼″ gate valve at 1.0′ 4, 1¼″ elbows at 2.0′	110.9	25.9	8.4	9.3
A–B	L	1¼	40	1, 1¼″ gate valve at 1.0′ 4, 1¼″ tees, run at 0.5′ 4, 1¼″ elbows at 2.0′	51.0	25.9	7.4	3.8
B–C	L	1	5	1, 1″ gate valve at 1.0′ 4, 1″ elbows at 1.0′	10.0	14.6	7.2	0.7
C–D	L	1	5	2, 1″ tees, run at 0.5′ 4, 1″ elbows at 1.0′	10.0	14.6	7.2	0.7
D–E	L	¾	10	1, ¾″ tee, run at 0.5′ 3, ¾″ elbows at 1.0′	13.5	9.7	12.0	1.6
E–F	L	¾	10	1, ¾″ tee, branch at 2′	12.0	8.7	10.0	1.2
F–G	L	¾	10	1, ¾″ globe valve at 20′ 1, ¾″ tee, run at 0.5′ 4, ¾″ elbows at 1.0′	34.5	5.0	3.7	1.3
Totals			180 (100%)		241.9 (134%)			18.6

Chlorides 3 ppm
Dissolved salts 40 ppm
Dissolved air Supersaturated
Ryznar index +12.1 at 60°F and +10.5 at 140°F

Sulfates 9 ppm
Free CO_2 2 ppm
Langelier index −2.5 at 60°F and −1.7 at 140°F

Records show no significant corrosion of copper by the water up to 150°F.

Copper water tube with wrought copper fittings has been selected, and is to be of type K for the water service and of type L for inside the building.

Water supply for the premises is to be metered at the point of entry by a disk-type meter. The system is to be of the upfeed riser type. A horizontal hot water storage tank is to provide hot water to the entire building, and is to be equipped with automatic tank control of water temperature set for 140°F.

The most extreme run of piping from the public main to the highest and most remote outlet is 137 ft in developed length, consisting of the following: 50 ft of water service, 12 ft of cold water piping from the water service valve to the individual branch to the horizontal hot water storage tank, and 75 ft therefrom to the top floor hot water outlet at the kitchen sink. Plans of the entire water supply system are available.

The building has a full cellar and three above-grade stories, each 10 ft (3 m) in height from floor to floor. The first floor is 2 ft (0.6 m) above curb level in front of the building. The public water main is located beneath the street, 5 ft (1.5 m) from the curb and 4 ft (1.2 m) below curb level.

On each of the above-grade stories, there are three dwelling units. Each dwelling unit has a kitchen containing a countertop sink and an automatic dishwasher, and a bathroom equipped with a close-coupled water closet and flush tank combination, a lavatory, and a bathtub with shower head above.

In the cellar are the following: a laundry room containing two automatic laundry washing machines and a slop sink, an employees' toilet room equipped with a lavatory and a close-coupled water closet and flush tank combination, and a porter's room in which a slop sink is provided.

On the building exterior are two hose bibs, one located at the front wall and another at the rear wall. They are remote from each other and are not to be used simultaneously for lawn watering.

Solution to Problem 2

Step 1 All information required for establishing a proper design basis has been obtained from appropriate sources.

Step 2 A schematic elevation of the building water supply system is provided in Fig. 3-2. This drawing was developed using the plans of the system. All piping connections have been shown in proper sequence as per the plans, and the developed lengths of each section of the basic design circuit have been determined therefrom. Fixtures and risers have been identified by combinations of letters and numbers, and those fixtures and branches having quick-closing valve outlets have been specially identified by means of an asterisk. All information required for establishing a proper design basis has been shown on the left side of the design sheet.

Step 3 For each section of the system, notations have been made showing the hot and cold water loads conveyed thereby in terms of water supply fixture units. Fixture unit values have been shown unenclosed by parentheses.

Step 4 Adjacent to fixture unit load notations, the demand in gallons per minute corresponding to such loads has been shown in parentheses. The demand in gallons per minute was determined from Table 3-2, applying the values shown therein under the heading, "Supply Systems Predominantly for Flush Tanks," in view of the fact that no flush valves are involved in this water supply system.

Step 5 The continuous demand posed by the two outside hose bibs, only one of which is to be used at a time, has been shown on the design sheet. This demand has been specially underlined and included in parentheses as gallons per minute, or otherwise designated CL, where added to fixture units of load. The normal demand posed by a hose bib was obtained from Table 3-1.

Step 6 All individual fixture supply pipes to water outlets have been sized on the design sheet in accordance with the minimum sizes shown under Standard Code Regulations in Table 3-15.

Step 7 All other parts of the system have been sized in accordance with the velocity limitations established for this system as the proper basis for design, i.e., 8 fps for all piping, except 4 fps for branches to quick-closing valves as noted by asterisks on the design sheet. Sizing was done in accordance with total water supply fixture units of load corresponding to total demand in each section. For those sections of the cold water header in the cellar which convey both demand of intermittently used fixtures and continuous demand of a hose bib, the total demand in gallons per minute was converted to equivalent water supply fixture units of load and proper sizes were determined therefor, although

DESIGN BASIS

PIPING:
Copper water tube with wrought copper fittings, type K for water service, type L for inside building.

PUBLIC SUPPLY SYSTEM:
10" public main located in street in front of building. Certified available pressure is: 55 psi maximum, and 40 psi minimum.

WATER CHARACTERISTICS:
Records show no significant corrosion of copper by the water up to 150°F.

WATER ANALYSIS:

pH	7.1
Hardness	23.00 ppm
Iron	0.07 ppm
Alkalinity	13.00 ppm
Chlorides	3.00 ppm
Sulphates	9.00 ppm
Dissolved salts	40.00 ppm
Free CO_2	2.00 ppm

Dissolved air—Supersaturated.

Langelier index:
−2.5 at 60°F; −1.7 at 140°F.

Ryznar index:
+12.1 at 60°F; +10.5 at 140°F.

ELEVATIONS:

Curb (as datum)	0.00'
Public main	−4.00
Cellar floor	−8.00
First floor	+2.00
Second floor	+12.00
Third floor	+22.00
Highest outlet	+25.00

HIGHEST OUTLET PRESSURE:
8 psi required.

DEVELOPED LENGTHS OF PIPING FROM PUBLIC MAIN TO HIGHEST AND FARTHEST OUTLET:
Sections: PM-A = 50'; A-B = 12'; B-C = 8'; C-D = 4'; D-E = 7'; E-F = 6.5'; F-G = 12.5'; G-H = 12'; H-I = 3'; I-J = 10'; J-K = 10'; K-L = 2'. TOTAL = 137'.

HOT WATER TEMPERATURE:
140°F System, tank control.

DESIGN VELOCITY LIMITS:
8 fps for all piping, except 4 fps for branches to quick-closing valves (noted by*).

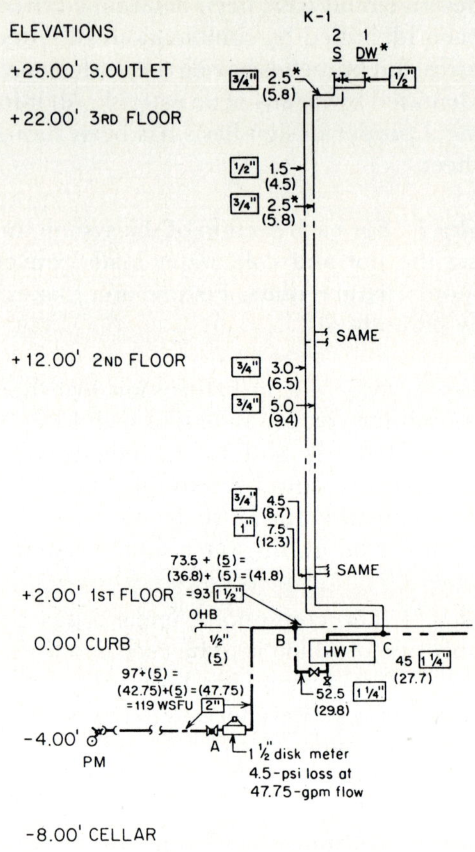

Fig. 3-2 Water supply system design sheet for Prob. 2.

KEY FOR NOTATIONS MADE IN SIZING PROCEDURE

Load in water supply fixture units is shown unenclosed.
Load in gpm of flow is shown in ( ).
Continuous load in gpm of flow is shown in (_).
Sizes selected are shown in ☐".

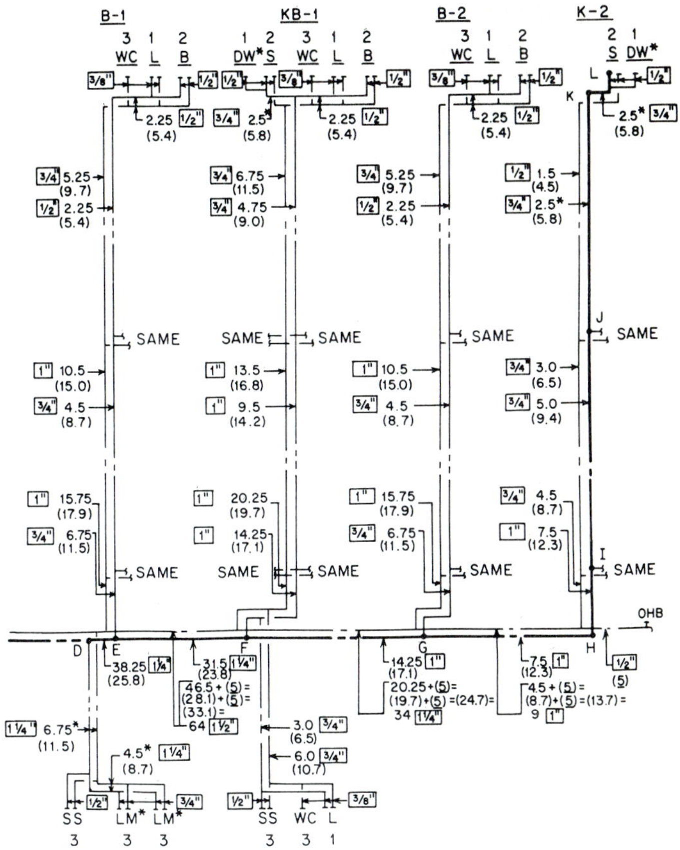

proper sizing could also have been done simply on the basis of demand rates in gallons per minute. Sizing was done using Tables 3-6*a* and *b* and 3-7*a* and *b*.

SUPPLEMENTARY CHECK RE PIPE FRICTION LOSSES IN BDC

To prove the adequacy of the simplified sizing method as applied in this second problem, a supplementary check has been made regarding the total pipe friction losses in the main lines and risers for the longest run of piping from the public water main to the highest and most remote water outlet. This run of piping was shown in heavy lines on the design sheet and letters marked thereon to denote the various sections of the run, the BDC.

The certified minimum pressure available at the public main was given as 40 psi. After appropriate deductions therefrom, the maximum pressure available for pipe friction losses during peak demand in the BDC was determined to be 14.2 psi. The appropriate deductions were as follows: 12.6 psi for static head loss due to rise from the public main to the highest outlet, 8.0 psi minimum pressure required at the top outlet, 4.5 psi friction loss through the 1½-in disk meter at 47.75 gpm flow rate, and 0.7 psi friction loss rated at 29.8 gpm flow rate through the horizontal hot water storage tank.

Table 3-10 shows the losses determined for each section of the circuit. Total pipe friction loss for the BDC was determined to be 11.3 psi, a total of 2.9 psi less than the 14.2 psi previously determined as the maximum pressure available for pipe friction losses during peak demand in the BDC. This proves that the sizes determined in this second problem on the basis of velocity limitation exclusively are adequate.

APPLICATION TO SYSTEMS IN HIGH BUILDINGS

This simplified method of sizing, based upon the velocity limitations which should be observed in design of building water supply systems, has much broader application than just to systems in one-, two-, and three-story buildings where ample excess pressure is available at the source of supply. These velocity limitations should be observed in all building water supply systems. Thus, the sizes determined by this method are the minimum sizes recommended for use in any case. Where pipe friction is an additional limiting factor to be considered in design, larger sizes may be required with respect to the sizing of main lines and risers.

DETAILED METHOD FOR SIZING SYSTEMS IN BUILDINGS OF ANY HEIGHT

For sizing water supply systems in buildings of any height, a detailed method has been developed in a step-by-step procedure which may be

Table 3-10
PIPE FRICTION LOSSES IN BASIC DESIGN CIRCUIT FOR PROB. 2
Copper Water Tube

Piping section	Tube type	Tube size, in	Developed length, ft	Equivalent length of fittings and valves, ft	Total equivalent length of section, ft	Flow rate q gpm	Friction (Charts 6-4*a*, 6-5*a*), psi/100′	Friction loss, psi
PM–A	K	2	50	1, 2″ tap at 2.0′ 1, 2″ gate valve at 2.0′ 4, 2″ elbows at 2.0′	62.0	47.75	2.9	1.8
A–B	L	2	12	3, 2″ elbows at 2.0′ 2, 2″ Ts, branch at 7′	32.0	47.75	2.6	0.83
B–C	L	1¼	8	1, 1¼″ gate valve at 1.0′ 5, 1¼″ elbows at 2.0′ 2, 1¼″ Ts, branch at 4′	27.0	29.8	9.5	2.57
C–D	L	1¼	4	1, 1¼″ T, run at 0.5′	4.5	27.7	8.4	0.38
D–E	L	1¼	7	1, 1¼″ T, run at 0.5′	7.5	25.8	7.2	0.54
E–F	L	1¼	6.5	1, 1¼″ T, run at 0.5′	7.0	23.8	6.4	0.45
F–G	L	1	12.5	1, 1″ T, run at 0.5′	13.0	17.1	9.6	1.25
G–H	L	1	12	1, 1″ elbow at 1.0′	13.0	12.3	5.4	0.70
H–I	L	1	3	1, 1″ T, run at 0.5′	3.5	12.3	5.4	0.19
I–J	L	¾	10	1, ¾″ T, run at 0.5′	10.5	9.4	11.5	1.21
J–K	L	¾	10	1, ¾″ elbow at 1.0′	11.0	5.8	4.8	0.53
K–L	L	¾	2	1, ¾″ T, run at 0.5′ 1, ¾″ globe valve at 15′ 2, ¾″ elbows at 0.5′	18.5	5.8	4.8	0.89
Totals			137 (100%)		209.5 (152.9%)			11.34

applied in the design of modern buildings. The procedure consists of 16 steps. They are as follows:

1. Obtain all information necessary for establishing a proper basis for sizing the system. Properness of the basis for sizing is contingent upon accuracy and reliability of the information applied. Such information should be obtained from responsible parties and appropriate local authorities recognized as sources of the necessary information. See the preceding discussion under "Preliminary Information."

2. Provide a schematic elevation of the complete water supply system. Show all piping connections in proper sequence and all fixture supplies. Identify all fixtures and risers by means of appropriate letters, numbers, or combinations thereof. Specially identify all piping conveying water at a temperature above 150°F (66°C), and all branch piping to such water outlets as automatic flush valves, solenoid valves, pneumatic valves, or quick-closing valves or faucets. Provide on the schematic elevation all the necessary information obtained as per step 1, and as discussed under "Preliminary Information."

3. Mark on the schematic elevation, for each section of the complete system, the hot and cold water loads conveyed thereby in terms of water supply fixture units in accordance with Table 3-14.

4. Mark on the schematic elevation, adjacent to all fixture unit notations, the demand in gallons per minute or liters per second corresponding to the various fixture unit loads in accordance with Table 3-2.

5. Mark on the schematic elevation, for appropriate sections of the system, the demand in gallons per minute or liters per second for outlets at which demand is deemed continuous, such as outlets for watering gardens, irrigating lawns, air-conditioning apparatus, refrigeration machines, and other similar equipment using water at a relatively continuous rate during peak demand periods. Add the continuous demand to the demand for intermittently used fixtures, and show the total demand at those sections where both types of demand occur.

6. Size all individual fixture supply pipes to water outlets in accordance with the minimum sizes permitted by regulations. Minimum fixture supply pipe sizes for common plumbing fixtures are given in Table 3-15.

7. Size all other parts of the water supply system in accordance with velocity limitations recognized as good engineering practice, with velocity limitations recommended by pipe manufacturers for avoiding accelerated deterioration and failure of their products under various conditions of service, and with velocity limitations generally recommended for minimizing the cost of pumping where water supply pumps are provided. [Sizing tables based on such velocity limitations and showing permissible loads in terms of water supply fixture units for each size and kind of piping material have been provided (Tables 3-3 through 3-8) and may be applied advantageously in this step.]
8. Assuming conditions of no flow in the system, calculate the amount of pressure available at the topmost fixture in excess of the minimum pressure required at such fixture for satisfactory supply conditions. The calculated excess pressure is the limit to which friction losses may be permitted for flow during peak demand in the system. (1 ft of water column = 0.433 psi pressure, and 1 m of water column = 9.795 kPa pressure.)
9. Determine which piping circuit of the system is the basic design circuit (BDC) for which pipe sizes in main lines and risers should be designed in accordance with friction loss limits. This circuit is the most extreme run of piping through which water flows from the public main, or other pressure source of supply, to the highest and most distant water outlet. The basic design circuit (BDC) should be specially identified on the schematic elevation of the system.
10. Mark on the schematic elevation the rated pressure loss due to friction corresponding to the demand through any water meter, water softener, or instantaneous or tankless hot water heating coil that may be provided in the basic design circuit.
11. Calculate the amount of pressure remaining and available for dissipation as friction loss during peak demand through the pipe, valves, and fittings in the basic design circuit. Deduct from the excess static pressure available at the topmost fixture (determined in step 8) the rated friction losses for any water meters, water softeners, or instantaneous or tankless hot water heating coils provided in the basic design circuit (determined in step 10).
12. Calculate the total equivalent length of the basic design circuit. Pipe sizes established on the basis of velocity limitations in step 7 for main lines and risers must be considered just tenta-

tive at this stage but may be deemed appropriate for determining corresponding equivalent lengths of fittings and valves in this step.

13. Calculate the permissible uniform pressure loss for friction in piping of the basic design circuit. The amount of pressure available for dissipation as friction loss due to pipe, fittings, and valves in the circuit (determined in step 11) should be divided by the total equivalent length of the circuit (determined in step 12). This establishes the pipe friction limit for the circuit in terms of pressure loss, in psi/ft (Pa/m) for the total equivalent pipe length. Multiply this value by 100 in order to express the pipe friction limit in terms of psi per 100 ft of length.

14. Set up a sizing table showing the rates of flow, for various sizes of the kind of piping to be used, corresponding to the permissible uniform pressure loss for pipe friction calculated for the basic design circuit (determined in step 13). Such rates may be determined from an accurate pipe friction chart appropriate for the kind of piping to be used and for the effects upon the piping of the quality of water to be conveyed thereby for extended service.

15. Size all parts of the basic design circuit, and all other main lines and risers which supply water upward to the highest water outlets on the system, in accordance with the sizing table set up in step 14. Where sizes determined in this step are larger than those previously established in step 7 (based just on velocity limitations), the increased size is applicable for limitation of friction.

16. Due consideration must be given to the action of the water on the interior of the piping, and proper allowance must be made where necessary as a design consideration, such as where the kind of piping selected and the characteristics of the water conveyed are such that an appreciable buildup of corrosion products or hard-water scale may be anticipated to cause a significant reduction in bore of the piping system and inadequate capacity for satisfactory supply conditions during the normal service life of the system. A reasonable allowance in such cases may be considered to be provision of at least one standard pipe size larger than the sizes determined in the preceding steps. Where the water supply is treated in such manner as to avoid buildup of corrosion products or hard-water scale, no allowance need be made in sizing piping conveying such treated water.

APPLICATION OF DETAILED METHOD TO ILLUSTRATIVE PROBLEMS

Problem 3

Draw a schematic elevation, and size the piping of the following water distributing system using the detailed sizing method:

A 102-family multiple dwelling, seven stories and basement in height, fronts on a public street and is to be supplied by direct street pressure from an 8-in public water main located beneath the street in front of the building. The public system is of cast iron and a hydrant flow test indicates a certified minimum available pressure of 75 psi. Top floor fixture outlets are 65 ft 8 in above the public main and require 8 psi flow pressure for satisfactory operation.

Authoritative water analysis reports show that the public water supply has a pH of 6.9, carbon dioxide content of 3 ppm, dissolved solids content of 40 ppm, and is supersaturated with air. Reports show that the public water supply has no significant corrosion effect on red brass for temperatures up to 150°F.

Cement-lined cast iron, class B, corporation water pipe, valves, and fittings have been selected for the water service pipe. Red brass pipe, standard pipe size, has been selected for the water distributing system inside the building.

Water supply for the building is to be metered at the point of entry by a compound meter installed in the basement. The system is to be of the upfeed riser type. A horizontal hot water storage tank is to provide hot water to the entire building, and is to be equipped with automatic tank control of water temperature set for 140°F. The tank is to have a submerged heat exchanger.

The most extreme run of piping from the public main to the highest and most remote outlet is 420 ft in developed length, consisting of the following: 83 ft of water service, 110 ft of cold water piping from the water service valve to the hot water storage tank, and 227 ft of hot water piping from the tank to the top floor hot water outlet at the kitchen sink. Plans of the entire water supply system are available.

The building has a basement and seven above-grade stories. The basement floor is 3 ft 8 in below curb level, the first floor is 5.0 ft above curb level, and the public water main is 5.0 ft below curb level. Each of the above-grade stories is 9 ft 4 in in height from floor to floor. The highest fixture outlet is 3 ft above floor level.

Fixtures provided on the system for the occupancies are as follows:

1. There are 17 dwelling units on each of the second, third, fourth, fifth, sixth, and seventh floors; and each dwelling unit is provided with a sink and domestic dishwashing machine in the kitchen,

and a close-coupled water closet and flush tank combination, a lavatory, and a bathtub with shower head above in a private bathroom.

2. The first floor is occupied for administrative and general purposes, and has the following provisions for such occupancy: one flush-valve supplied water closet and one lavatory in an office toilet room; one flush-valve supplied water closet, one flush-valve supplied urinal and one lavatory in a men's toilet room; two flush-valve supplied water closets and one lavatory in each of two women's toilet rooms; a sink and domestic dishwashing machine in a demonstration kitchen; one sink in an office kitchen; one sink in a craft room; and two drinking fountains in the public hall.

3. The basement is occupied for building equipment rooms, storage, utility, laundry, and general purposes and has the following provisions for such occupancy: one flush-valve supplied water closet and one lavatory in a women's toilet room; one flush-valve supplied water closet, one lavatory, and one shower stall in a men's toilet room; one service sink and six automatic laundry washing machines in a general laundry room; one faucet above a floor drain in the boiler room; and one valve-controlled primary water supply connection to the building heating system.

4. At each story and in the basement, a service sink is provided in a janitor's closet in the public hall.

5. Four outside hose bibs (only two to be used at any time) are provided for lawn watering at appropriate locations on the exterior of the building.

Fixture arrangements are typical on the six upper floors of the building, and 24 sets of risers are provided. Of these, 5 sets are for back-to-back bathrooms, 2 sets are for back-to-back kitchens, 4 sets are for back-to-back kitchen and bathroom groups, 9 sets are for separate kitchens, 3 sets are for separate bathrooms, and one set is for a service sink on each floor above the basement. Fixtures on the first floor are connected to adjacent risers. Basement fixtures are connected to overhead mains, which also supply directly the four outside hose bibs.

Solution to Problem 3

Step 1 All information required for establishing a proper design basis has been obtained from appropriate sources.

Step 2 A schematic elevation of the building water supply system is provided in Fig. 3-3. This drawing was developed using the plans of the system. All piping connections have been shown in proper sequence as per the plans, and the developed lengths of each section of the basic design circuit have been determined therefrom. Fixtures and risers have been identified by combinations of letters and numbers, and those fixtures and branches having quick-closing valve outlets have been specially identified by means of an asterisk. All information required for establishing a proper design basis has been shown on the left side of the design sheet.

Step 3 For each section of the system, notations have been made showing the hot and cold water loads conveyed thereby in terms of water supply fixture units. Fixture unit values have been shown unenclosed by parentheses.

Step 4 Adjacent to fixture unit load notations, the demand in gallons per minute corresponding to such loads has been shown in parentheses. The demand in gallons per minute was determined from Table 3-2, applying the values shown therein under the heading, "Supply Systems Predominantly for Flush Tanks," for all piping except for short branch piping which supplies water to water closets and urinals equipped with flush valves on the first floor and in the basement.

Step 5 The continuous demand posed by the four outside hose bibs, only two of which are to be used at any time, has been shown on the design sheet. This demand has been specially underlined and included in parentheses as gallons per minute, or otherwise designated CL, where added to fixture units of load. The normal demand posed by a hose bib was obtained from Table 3-1.

Step 6 All individual fixture supply pipes to water outlets have been sized on the design sheet in accordance with the minimum sizes shown under Standard Code Regulations in Table 3-15.

Step 7 All other parts of the system have been sized in this step in accordance with the velocity limitations established for this system as the proper basis for design, i.e., 8 fps for all piping, except 4 fps for branches to quick-closing valves as noted by asterisks on the design sheet. Sizing was done in accordance with total fixture units of load corresponding to total demand in each section. For those sections of the cold water header in the basement which convey both demand of intermittently used fixtures and continuous demand of hose bibs, the

DESIGN BASIS

PIPING:
Cement-lined cast iron, class B, corporation water pipe, valves and fittings for water service pipe. Red brass pipe, standard pipe size, for water distributing system within the building.

PUBLIC SUPPLY SYSTEM:
8" cast iron public water main located in street in front of building. Hydrant flow test indicates a certified minimum available pressure of 75 psi.

WATER CHARACTERISTICS:
Authoritative reports show that the public water supply has no significant corrosion effect on red brass for temperatures up 150°F.

WATER ANALYSIS:

pH	6.9
CO_2	3.0 ppm
Dissolved solids	40.0 ppm
Dissolved air – saturated.	

ELEVATIONS:

Curb (as datum)	0.00'
Public main	-5.00'
Basement floor	-3.67'
First floor	+5.00'
Second floor	+14.33'
Third floor	+23.00'
Fourth floor	+31.67'
Fifth floor	+40.33'
Sixth floor	+49.00'
Seventh floor	+57.67'
Highest outlet	+60.67'

HIGHEST OUTLET PRESSURE:
8 psi required.

DEVELOPED LENGTHS OF PIPING FROM PUBLIC MAIN TO HIGHEST AND FARTHEST OUTLET:
Sections: PM-A = 83'; A-B = 30'; B-C = 10'; C-D = 10'; D-E = 10'; E-F = 10'; F-G = 20'; G-H = 30'; H-I = 15'; I-J = 15'; J-K = 15'; K-L = 12'; L-M = 12'; M-N = 12'; N-O = 12'; O-P = 12'; P-Q = 12'; Q-R = 12'; R-S = 12'; S-T = 12'; T-U = 16.65'; U-V = 8.67' V-W = 8.67'; W-X = 8.67'; X-Y = 8.67'; Y-Z = 12.67'
Total developed length of basic design circuit = 420'.

HOT WATER TEMPERATURE:
140°F system, tank control.

DESIGN VELOCITY LIMITS:
8 fps for all piping, except 4 fps for branches to quick-closing valves (noted by*).

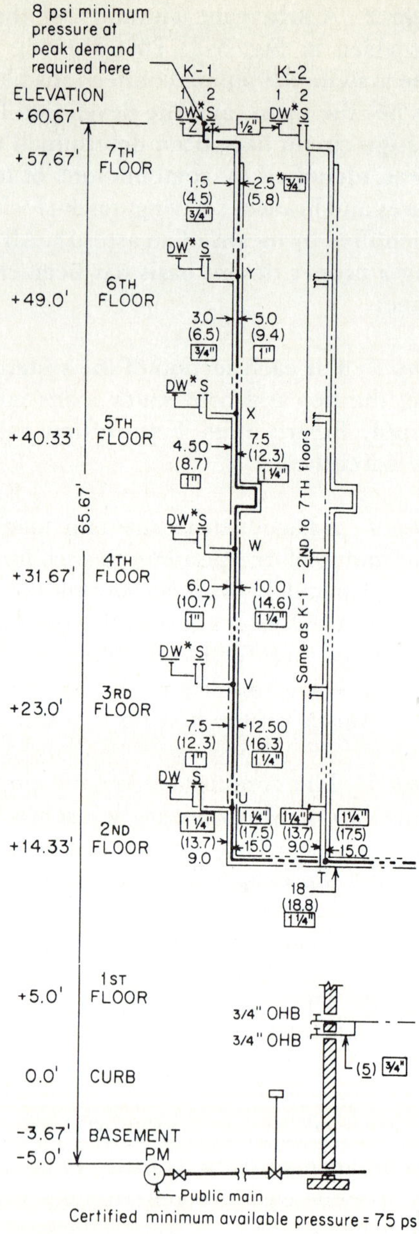

Fig. 3-3 Water supply system design sheet for Prob. 3.

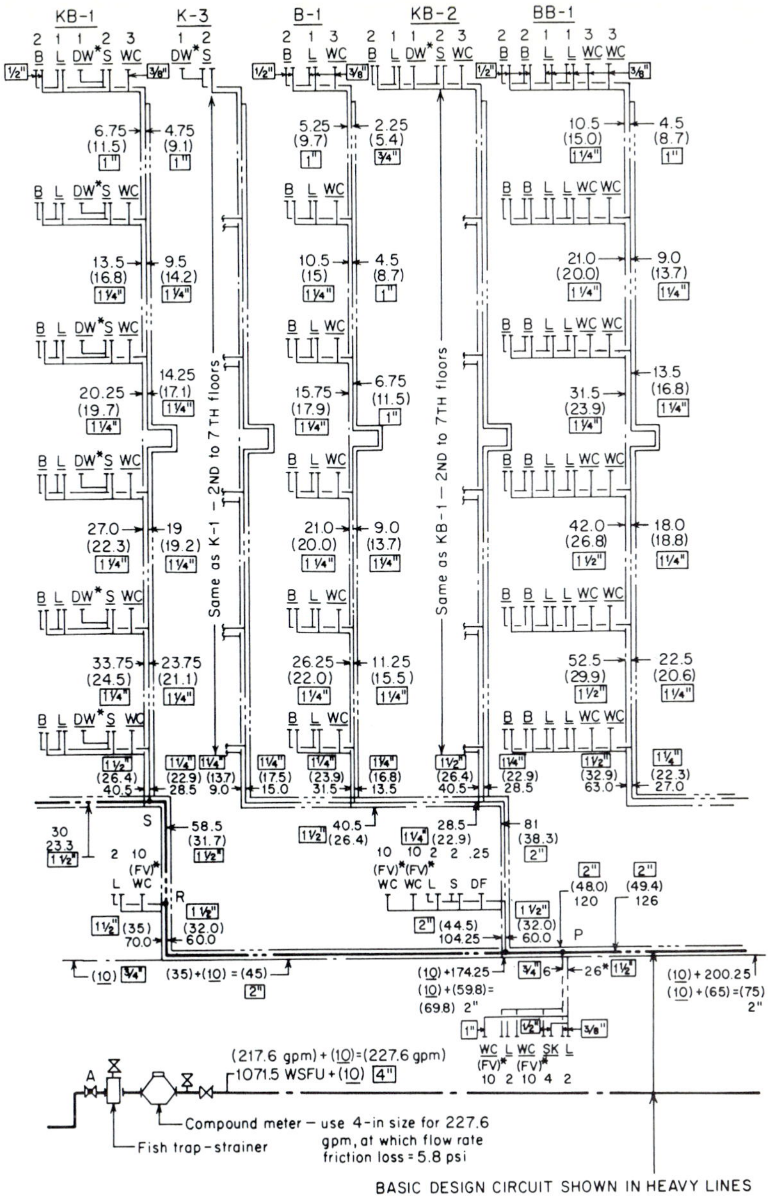
KB-1
K-3
B-1
KB-2
BB-1
Same as K-1 — 2ND to 7TH floors
Same as KB-1 — 2ND to 7TH floors
(217.6 gpm) + (10) = (227.6 gpm)
1071.5 WSFU + (10) 4"
Compound meter — use 4-in size for 227.6 gpm, at which flow rate friction loss = 5.8 psi
Fish trap-strainer
BASIC DESIGN CIRCUIT SHOWN IN HEAVY LINES

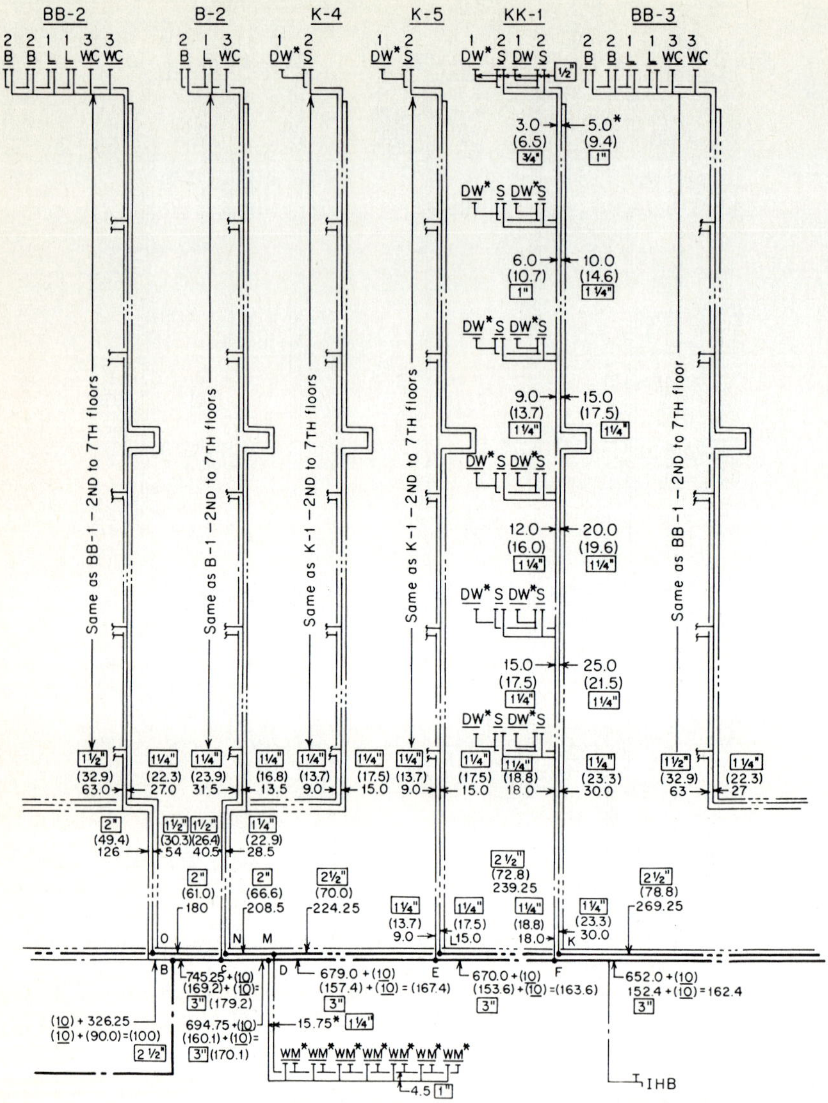

Fig. 3-3 (***Continued***)

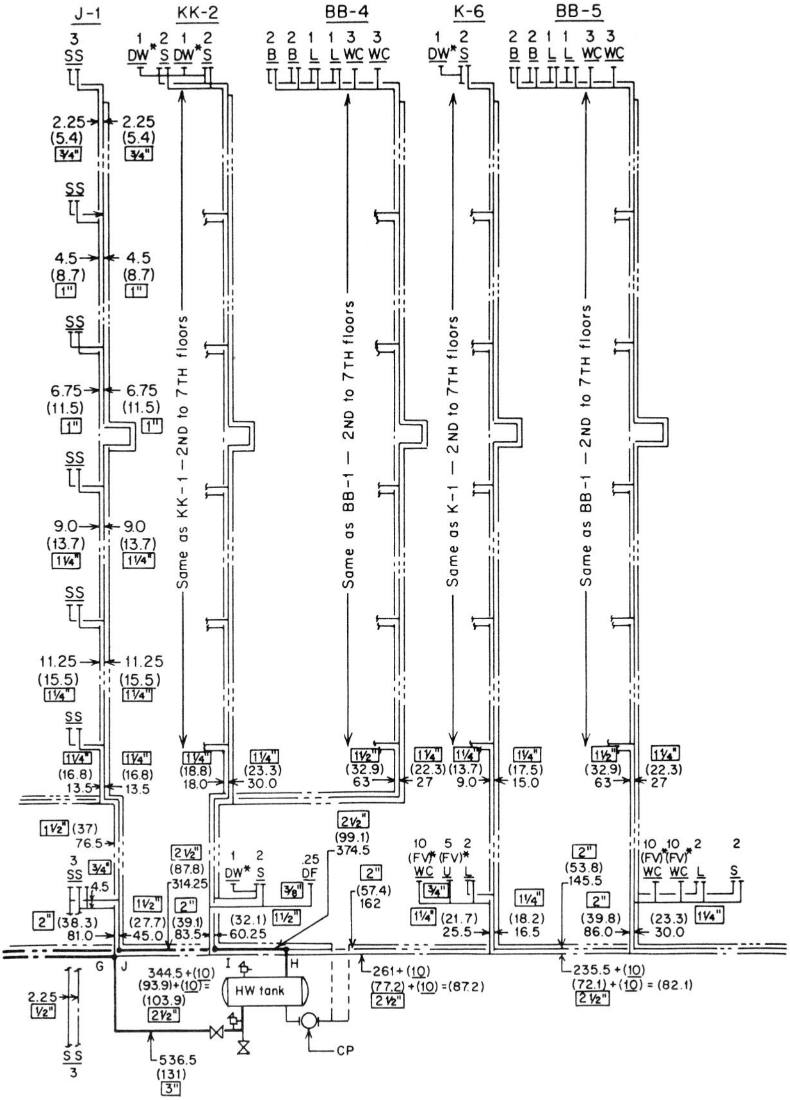
J-1
KK-2
BB-4
K-6
BB-5
Same as KK-1 —2ND to 7TH floors
Same as BB-1 — 2ND to 7TH floors
Same as K-1 — 2ND to 7TH floors
Same as BB-1 —2ND to 7TH floors
HW tank
CP
G
J
I
H
536.5
(131)
3"
344.5 + (10)
(93.9) + (10) =
(103.9)
2½"
261 + (10)
(77.2) + (10) = (87.2)
2½"
235.5 + (10)
(72.1) + (10) = (82.1)
2½"

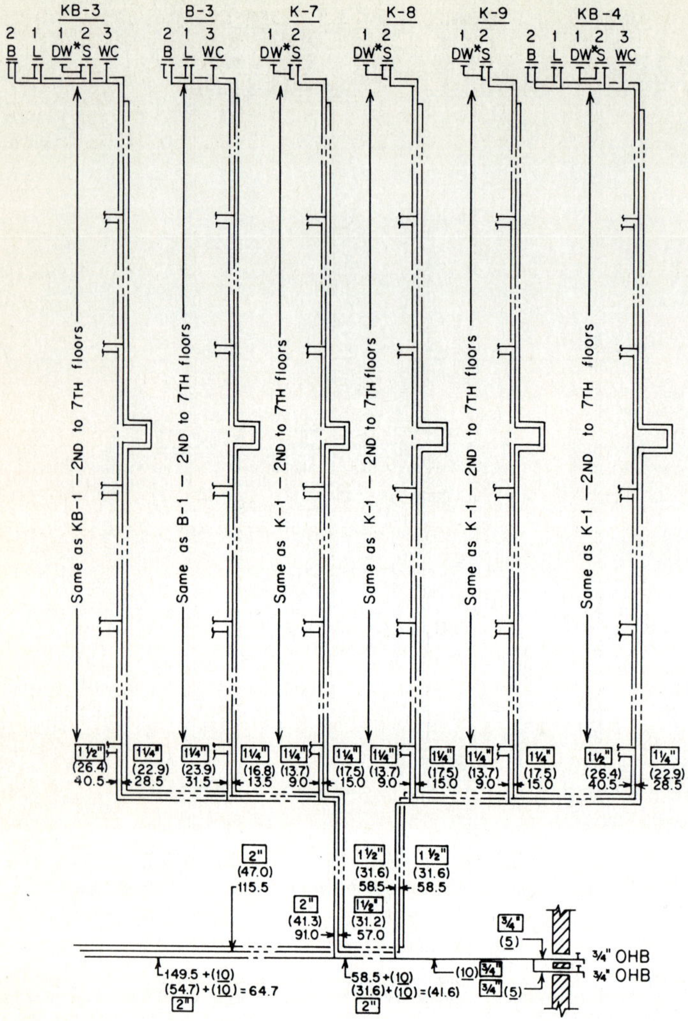

Fig. 3-3 (*Concluded*)

total demand in gallons per minute was converted to equivalent water supply fixture units of load and proper sizes were determined therefor, although proper sizing could also have been done simply on the basis of demand rates in gallons per minutes. Sizing was done using Tables 3-3 through 3-8 specifically Tables 3-4*a* and 3-4*b*, for sizing piping inside the building.

Step 8 Assuming conditions of no flow in the system, the amount of excess pressure available at the topmost fixture in excess of the minimum pressure required at the fixture for satisfactory supply conditions was determined as follows:

$$\text{Excess pressure available} = 75 \text{ psi} - 8 \text{ psi} - (65.67 \times 0.433)$$
$$= 38.6 \text{ psi}$$

Step 9 The basic design circuit of the water supply system for the building was specially identified and shown in heavy lines on the schematic elevation provided in step 2. For each of the 26 sections of the circuit,

Table 3-11
PRESSURE CALCULATIONS FOR BASIC DESIGN CIRCUIT

Minimum at public main	75.0 psi
Loss in rise to top outlet (65.67 ft × 0.433)	−28.4 psi
Static pressure at top outlet	46.6 psi
Minimum pressure at top outlet	− 8.0 psi
Excess static pressure at top outlet available for friction loss	38.6 psi
Friction loss through 4-in compound meter at 227 gpm flow rate (manufacturer's charts)	− 5.8 psi
	32.8 psi
Friction loss through horizontal hot water storage tank assumed for rated flow at 8 fps	− 0.7 psi
Maximum pressure remaining for friction in pipe, valves, and fittings	32.1 psi
Developed length of circuit from public main to top outlet	420 ft
Equivalent length for valves and fittings in circuit (based on sizes established on velocity limitation basis)	363 ft
Total equivalent length of circuit	783 ft
Maximum uniform pressure loss for friction in basic design circuit (32.1 psi/783 ft)	0.04 psi/ft or 4.0 psi/100 ft

the developed length was given on the design sheet as determined from the plans of the system.

Step 10 The rated pressure loss through the compound water meter selected for this system was determined from appropriate meter data to be 5.8 psi for peak demand flow rate of 227.6 gpm. This has been noted on the design sheet. The rated pressure loss for flow through the horizontal hot water storage tank, i.e., entrance and exit losses, may be assumed to be approximately 1.6 ft head or 0.7 psi.

Step 11 The amount of pressure available for dissipation as friction loss during peak demand through pipe, valves, and fittings in the basic design circuit is

$$38.6 - 5.8 - 0.7 = 32.1 \text{ psi}$$

Step 12 In step 7, tentative pipe sizes for the main lines and risers were established based on velocity limitations. Using such tentative sizes for the basic design circuit, corresponding equivalent lengths for valves and fittings were determined and added to the developed length to calculate the total equivalent length of the circuit. The equivalent length for valves and fittings was found to be 363.2 ft, which when added to the 420 ft developed length resulted in a total equivalent length for the basic design circuit of 783.2 ft.

Step 13 The maximum uniform pressure loss for friction in the basic design circuit is

$$32.1 \text{ psi}/783.2 \text{ ft} = 0.04 \text{ psi/ft or } 4.0 \text{ psi}/100 \text{ ft}$$

This is the pipe friction limit for the basic design circuit. It is to be applied for sizing all the main lines and risers supplying water to fixtures on upper floors of the building.

Step 14 In Table 3-12 flow rates have been tabulated, through various sizes of red brass pipe of standard pipe size, that correspond to the velocity limits of 4 and 8 fps, and to the friction limit of 4.0 psi/100 ft of total equivalent piping length. The values shown therein for velocity limitations were taken from the tables cited in step 7. The values shown therein for friction limitation were taken directly from Chart 3-2*a*, one of the accurate pipe friction charts presented earlier in this chapter. The chart applied to red brass pipe of standard pipe size and was appropriate in view of the water supply conditions and surface condition, "fairly smooth."

Table 3-12
SIZING TABLE FOR SYSTEM IN PROB. 3
Red Brass Pipe, Standard Pipe Size

Nominal pipe size, in	Velocity limit flow rate at $V=4$ fps		Velocity limit flow rate at $V=8$ fps		Friction limit flow rate at 4.0 psi/100 ft, gpm
	WSFU (col. A)	gpm	WSFU (col. A)	gpm	
½	1.5	3.8	3.7	7.6	2.8
¾	3.0	6.6	8.4	13.2	5.8
1	6.3	11.1	26.4	22.0	11.7
1¼	16.8	18.3	75.0	36.6	22.5
1½	36.3	25.2	130.0	50.4	33.0
2	92.0	41.6	291.0	83.2	66.0
2½	181.0	61.2	492.0	122.4	112.0
3	335.0	92.0	842.0	184.0	288.0
4	685.0	158.0	1920.0	316.0	380.0

Note: Apply the column headed "Velocity limit, $V=4$ fps," to size branches to quick-closing valves. Apply the column headed "Velocity limit, $V=8$ fps," to all piping other than individual fixture supplies. Apply the column headed "Friction limit," just for sizing piping that conveys water to top floor outlets. Where two columns apply and two different sizes are indicated, select the larger size.

Step 15 All the main lines and risers on the design sheet have been subjected to sizing in accordance with the friction limitation for the basic design circuit. Where sizes determined in this step were larger than those previously determined in step 7 (based on velocity limitation), the increased size was noted directly on the design sheet. Increased sizes were made in all risers and in some parts of the main lines in this system. As an example, in the basic design circuit the sizes of many sections were increased and may be specifically cited as follows: sections J–K, K–L, and L–M were increased from 2-in to 2½-in; sections O–P and P–Q were increased from 1½-in to 2-in; sections Q–R, R–S, and S–T were increased from 1¼-in to 1½-in; sections T–U, U–V, and V–W were increased from 1-in to 1¼-in; section W–X was increased from ¾-in to 1¼-in; and section X–Y was increased from ¾-in to 1-in.

Step 16 From the characteristics of the water supply stated in the problem, it is recognized that the water is relatively noncorrosive and nonscaling. Consequently, there is no need for additional allowance in sizing in this case.

Problem 4

Provide a sizing table showing flow rates corresponding to the appropriate velocity and friction limits for the following water distributing system:

An office building, 20 stories in height, is designed to be supplied with water by means of two gravity water supply tanks elevated at the same height above the building roof. The tanks are to supply a standpipe system through connections at the tank bottoms and the building's potable water distributing system through connections 2 ft 3 in above the tank bottoms. Three pumps, one of which is reserved for standby service, draw water under direct pressure from two public water mains and, at the additional head required, supply the tanks at a rate double the peak demand on the system. The tanks and pumping equipment are designed to maintain as a normal operating condition a minimum tank water line 41 ft above the highest fixture outlets at the top story. The minimum pressure required at these outlets is 8 psi.

Authoritative water analysis reports show that the public water supply has a pH of 7.0, carbon dioxide content of 4 ppm, dissolved solids content of 55 ppm, and is supersaturated with air. Records show no significant corrosion of copper by the water up to 150°F.

Copper water tube, type L, with wrought copper fittings has been selected for this system. Joints are to be made with solder consisting of 95 percent tin and 5 percent antimony.

A horizontal hot water storage tank with submerged heater is provided in the cellar and is to be equipped with an automatic tank control set for 140°F hot water temperature.

The most extreme run of piping measured from the water supply tank connection to the highest and most remote outlet is 600 ft in developed length, consisting of the following: 210 ft of downfeed cold water riser from the supply tank connections down to the cellar, 75 ft of cold water piping to the horizontal hot water storage tank in the cellar, 140 ft of hot water piping from the tank to the base of the most extreme riser in the cellar, and 175 ft of hot water riser piping to the highest hot water outlet.

Fixtures are provided on each floor for the occupancies. Toilet rooms for men and for women are provided on each floor. Water closets, urinals, and lavatories are provided in men's toilet rooms, and water closets and lavatories are provided in women's toilet rooms. Drinking fountains are provided on each floor at appropriate locations. Janitor's closets on each floor contain mop sinks. All water closets, urinals, lavatories, and drinking fountains are supplied by means of quick-closing devices, including automatic flush valves, self-closing faucets, and push-button quick-closing valves.

Solution to Problem 4 (Steps 8 through 16)

Step 8 Assuming conditions of no flow in the system, the amount of excess pressure available at the topmost fixture in excess of the minimum pressure required at the fixture for satisfactory supply conditions is as follows:

$$\text{Excess pressure available} = (41 \text{ ft} \times 0.433 \text{ psi/ft}) - 8 \text{ psi}$$
$$= 17.75 - 8 = 9.75 \text{ psi}$$

Step 9 The problem states that the most extreme run of piping measured from the water supply tank connection to the highest and most remote outlet is 600 ft in developed length. This is the "basic design circuit' of the system.

Step 10 The rated pressure loss for flow through the horizontal hot water storage tank, i.e., entrance and exit losses, may be assumed for flow at 8 fps to be approximately 1.6 ft head or 0.7 psi.

Step 11 The amount of pressure available for dissipation as friction loss during peak demand through pipe, valves, and fittings in the basic design circuit is

$$9.75 \text{ psi} - 0.7 \text{ psi} = 9.05 \text{ psi}$$

Step 12 Total equivalent length of the basic design circuit, established tentatively based upon its developed length plus 50 percent as an appropriate assumed allowance for equivalent length of copper water tube fittings and valves, is as follows:

$$600 + (0.50 \times 600) = 600 + 300$$
$$= 900 \text{ ft}$$

Step 13 The maximum permissible uniform pressure loss for friction in the basic design circuit is

$$9.05 \text{ psi}/900 \text{ ft} = 0.01006 \text{ psi/ft or } 1.0 \text{ psi}/100 \text{ ft}$$

Step 14 In Table 3-13 flow rates have been tabulated, through various sizes of copper water tube, type L, that correspond to the velocity limit of 4 fps, and to the friction limit of 1.0 psi/100 ft of total equivalent piping length. The values shown therein for the velocity limit of 4 fps were taken from the tables cited in step 7 of the detailed method. The values shown therein for the friction limit of 1.0 psi/100 ft were taken directly from Chart 3-5*a*, one of the accurate pipe friction charts pre-

Table 3-13
SIZING TABLE FOR SYSTEM IN PROB. 4
Copper Water Tube, Type L

Nominal pipe size, in	Velocity limit flow rate at $V = 4$ fps		Friction limit flow rate at 1.0 psi/100 ft, gpm
	WSFU (col. B)	gpm	
¾		6.0	2.4
1		10.3	4.8
1¼	5.0	15.7	8.6
1½	8.0	22.8	13.5
2	26.0	38.6	28.0
2½	75.0	59.5	49.0
3	178.0	85.0	80.0
4	544.0	149.0	170.0

Note: Apply the column headed "Velocity limit" to size all branches from main lines and risers. Apply the column headed "Friction limit" to size main lines and risers supplying upper floor outlets.

sented earlier in this chapter. This chart was appropriate in view of the water supply conditions and the surface condition "fairly smooth." All branches from main lines and risers should be sized in accordance with the velocity limit. All main lines and risers should be sized in accordance with the friction limit.

STANDARD CODE REGULATIONS

Water Service Pipe The water service pipe shall be of sufficient size to furnish an adequate supply of water to meet the requirements of the building at peak demand. No water service pipe shall be smaller than ¾ in nominal size, except that where direct supply flush valves are supplied at pressure from the water service pipe, such pipe shall be at least 1¼ in nominal size.

Minimum Available Pressure The distributing system shall be designed on the basis of the minimum available pressure at the water main or other source of water supply.

Demand Load The water supply demand load in the distributing system shall be based upon the number, type, and probable simultaneous use of the fixtures to be supplied.

Table 3-14
DEMAND LOAD OF FIXTURES

Fixture	Occupancy	Type of supply control	Load values assigned, water supply fixture units		
			Cold	Hot	Total
Water closet	Public	Flush valve	10		10
Water closet	Public	Flush tank	5		5
Urinal	Public	1″ (25.4 mm) flush valve	10		10
Urinal	Public	¾″ (19 mm) flush valve	5		5
Urinal	Public	Flush tank	3		3
Lavatory	Public	Faucet	1.5	1.5	2
Bathtub	Public	Faucet	3	3	4
Showerhead	Public	Mixing valve	3	3	4
Service sink	Offices, etc.	Faucet	2.25	2.25	3
Kitchen sink	Hotel, restaurant	Faucet	3	3	4
Drinking fountain	Offices, etc.	⅜″ (9.52 mm) valve	0.25		0.25
Water closet	Private	Flush valve	6		6
Water closet	Private	Flush tank	3		3
Lavatory	Private	Faucet	0.75	0.75	1
Bathtub	Private	Faucet	1.5	1.5	2
Shower stall	Private	Mixing valve	1.5	1.5	2
Kitchen sink	Private	Faucet	1.5	1.5	2
Laundry trays (1 to 3)	Private	Faucet	2.25	2.25	3
Combination fixture	Private	Faucet	2.25	2.25	3
Dishwashing machine	Private	Automatic		1	1
Laundry machine [8 lb (3.6 kg)]	Private	Automatic	1.5	1.5	2
Laundry machine [8 lb (3.6 kg)]	Public or general	Automatic	2.25	2.25	3
Laundry machine [16 lb (7.3 kg)]	Public or general	Automatic	3	3	4

Note: For fixtures not listed, loads should be assumed by comparing the fixture with one listed using water in similar quantities and at similar rates. The assigned loads for fixtures with both hot and cold water supplies are given for separate hot and cold water loads and for total load, the separate hot and cold water loads being three-fourths of the total load for the fixture in each case.

Sizing Procedure The procedure for sizing the piping of the water supply distributing system shall conform to good engineering practice. Adequate and approved design factors shall be used to determine pipe sizes.

The size of piping, other than individual fixture supply pipes, should be such that the velocity of water flow during maximum demand will not exceed 8 fps (2.4 m/s) for piping aboveground, within dwelling portions of buildings.

In determining the available pressure at outlets, proper allowance should be made for the pressure loss owing to friction in the piping.

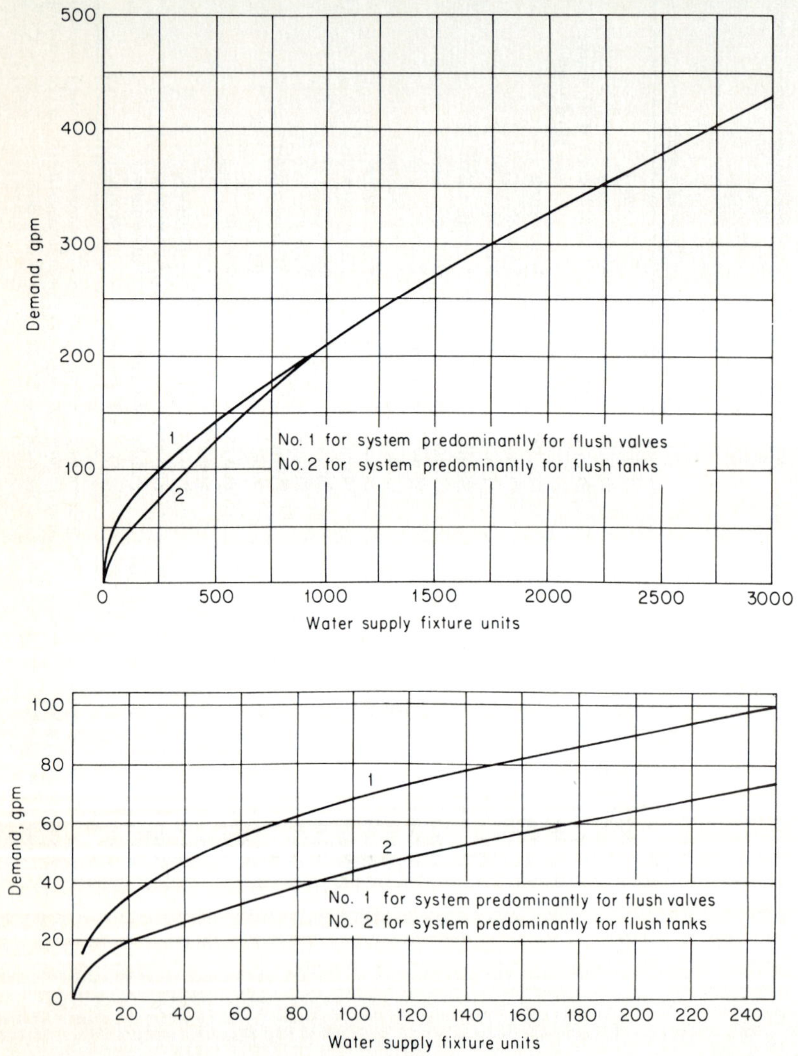

Fig. 3-4 Curves for estimating demand load.

Estimated Water Supply Demand Load For purposes of estimating the water supply load, the demand load values, in terms of water supply fixture units, for different plumbing fixtures under several conditions of service, are given in Table 3-14.

The estimated demand load for fixtures used intermittently on any supply pipe, in gallons per minute corresponding to the total number of water supply fixture units, is given in two charts in Fig. 3-4.

To estimate the total demand in gallons per minute, the demands for outlets such as hose connections and air-conditioning apparatus, which impose continuous demand during periods of heavy use, should be calculated separately and added to the demand for fixtures used intermittently.

Fixture Supply Pipes The minimum size of fixture supply pipes shall be as shown in Table 3-15. The fixture supply pipe shall terminate not more than 30 in (762 mm) from the point of connection to the fixture and in every instance shall extend into the room through the floor or wall adjacent to the fixture.

Risers The minimum size of water supply risers shall be not less than ½ in (12.7 mm) when no flush valves are supplied directly therefrom, nor less than

Table 3-15
MINIMUM SIZE OF FIXTURE SUPPLY PIPES

Fixture or device	Size in	Size mm
Bathtub	½	12.7
Combination sink and laundry tray	½	12.7
Drinking fountain	⅜	9.5
Dishwashing machine (domestic)	½	12.7
Kitchen sink (domestic)	½	12.7
Kitchen sink (commercial)	¾	19.0
Lavatory	⅜	9.5
Laundry tray (1, 2, or 3 compartments)	½	12.7
Shower (single head)	½	12.7
Sink (service, slop)	½	12.7
Sink (flushing rim)	¾	19.0
Urinal [1″ (25.4 mm) flush valve]	1	25.4
Urinal [¾″ (19.0 mm) flush valve]	¾	19.0
Urinal (flush tank)	½	12.7
Water closet (flush tank)	⅜	9.5
Water closet (flush valve)	1	25.4
Hose bib	½	12.7
Wall hydrant or sill cock	½	12.7

Note: For fixtures not listed in the above table, the minimum size of fixture supply pipes shall be the same as given in the table for comparable fixtures.

1¼ in (31.7 mm) when one or two flush valves are supplied directly therefrom, nor less than 1½ in (38.1 mm) when three or more flush valves are supplied directly therefrom.

Allowance for Water Characteristics Where the piping is subject to excessive corrosion or deposits because of characteristics of the water, pipe sizes shall be at least one standard pipe size larger than the minimums specified in this section.

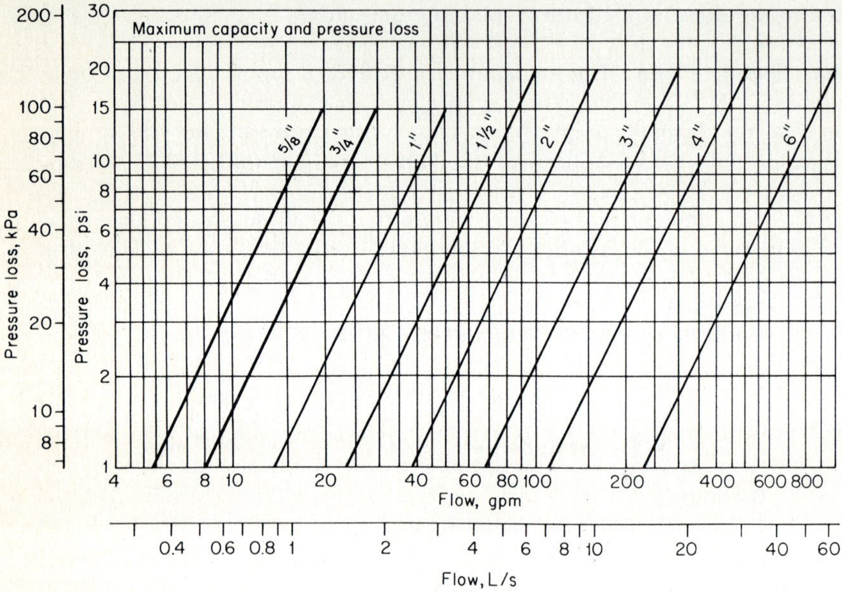

Fig. 3-5 Pressure loss in cold water meters, displacement type.

Section **4**

Hot Water Supply Systems

Hot water is essential for any building occupied by human beings. And a variety of other buildings, factories, animal shelters, warehouses, etc., also require hot water for various uses.

Since heat is required to raise the temperature of incoming cold water to convert it to hot water, energy conservation is always a consideration in system design. Yet this aspect of design is often neglected when time shortages require a rapid completion of a project.

Further, excessive water temperature or pressure can lead to serious accidents among system users. For this reason, extreme care is needed in the design of every hot water system.

This section covers the many factors in the safe, economic design of building hot water systems. Using the data provided, a designer can produce a safe system that will operate reliably and economically for many years.

OBJECTIVES OF STANDARD DESIGN

Standard design of hot water supply systems has the following objectives: (1) compliance with applicable regulations; (2) safe, satisfactory quality of service performance; (3) efficient utilization of available, economical sources of heat; (4) economy and durability of installation; and (5) economy and convenience of operation and maintenance of the completed installation.

In the latter portion of this chapter are presented provisions dealing with hot water supply systems which have become relatively common in standard code regulations throughout the United States. Very often these regulations are not found in just a single code, but rather in several different codes or sets of ordinances. They may be found scattered frequently amongst the provisions of building codes, plumbing codes, and sanitary or health codes. However, all applicable regulations should be observed in designing hot water supply systems.

The service performance of systems should be both safe and of satisfactory quality. Safety must be built into the systems by provisions to protect against the hazards of excessive pressure and excessive temperature so as to prevent injury to persons and damage to property which might otherwise occur. Satisfactory quality of service performance may be attained by designing the system so as to satisfy the demand for hot water at properly controlled volume, pressure, and temperature conditions.

Hot water temperature should be suitable for the service conditions required for safe use and proper function at fixtures. For most hot water supplied fixtures, at which persons come in contact with the water, the hot water temperature should be limited to a range of 130 to 140°F.

DEMAND AND HEATER CAPACITIES

The demand for water at 180°F (82°C) temperature is posed almost exclusively by the high-temperature water requirements for dishwashing machines and dishwashing fixtures in commercial kitchens serving employee dining spaces or public restaurants. Standard regulations require that, where food or drink is served to the public or to employees and the dishes, glasses, or cutlery are to be reused, such articles shall be washed effectively and then sanitized before reuse. The water temperature required for sanitizing purposes in this type of equipment is generally recognized as being 180°F (82°C). As a result of being sanitized by water at this high temperature, washed articles become air-dried in just a few moments after removal from the machine. This eliminates any need for drying articles by other methods.

Where glasses and cutlery are washed and sanitized in individual glass washing and silver washing machines or fixtures, such equipment should be understood to require 180°F (82°C) water for proper performance.

In most cases where dishes also are to be washed, there is no special need for separate glass washing and silver washing equipment, for dishwashing machines generally may be used efficiently for washing such items.

The only plumbing fixtures which may be considered suitable for using hot water at 165°F (74°C) safely are those of the machine type having enclosures which protect persons against being scalded during use of the fixture. These fixtures are known as automatic domestic dishwashing machines and automatic domestic laundry washing machines. Although they may perform satisfactorily in most cases with hot water at 140°F (60°C), many manufacturers of such machines have recommended that they be supplied with hot water at 165°F (74°C) for best performance.

To accurately estimate the demand for 180°F (82°C) water for any given installation, the amount of dishwashing to be performed should be determined first, and then suitable equipment and machines should be selected to match the dishwashing needs. Thereafter, the demand may be estimated directly from the data and recommendations contained in specification sheets available from dishwashing machine manufacturers regarding the performance of, and the water requirements for, their products. The rated 180°F (82°C) water requirements for the machines may be used to estimate the maximum instantaneous demand and also the maximum total daily demand.

The demand for 165°F (74°C) water supply to automatic domestic dishwashing machines and laundry washing machines should be estimated from data and recommendations contained in the specification sheets available from the manufacturer in each case. For different sizes and models of machines, there may be considerable variation in the water supply requirements.

Except for the high-temperature water requirements posed by dishwashing machines and dishwashing fixtures in commercial kitchens and for the slightly lower temperature hot water recommendations of manufacturers for automatic domestic dishwashing and laundry washing machines, the demand for hot water supply at ordinary plumbing fixtures in buildings is for water at 140°F (60°C). This is recognized as being the most desirable temperature for supplying hot water to plumbing fixtures as it is sufficiently high so that normal washing tasks may be accomplished and sufficiently low so that scalding of persons may be prevented and excessive corrosion and expansion effects in the system may be avoided.

The demand for 140°F (60°C) water in buildings depends upon numerous factors and conditions. They include (1) the occupancy classification of the building or portion supplied, (2) the number of persons accommodated, (3) the number and kind of plumbing fixtures, (4) any special equipment installed for occupancies, (5) the time of the day,

(6) the season of the year, (7) the installation of water conservation devices, and (8) whether hot water is supplied to occupancies at no extra charge.

Demand is never absolutely constant throughout the day. It is subject to extreme limits of variation. An objective for design of water-heating installations is to satisfy efficiently the peak demand for hot water. This involves two essential considerations: the peak demand rate and the duration of the peak draw period. They form the basis upon which to determine the most suitable capacities of heaters and storage tanks, the arrangement of water-heating equipment, and the sizes of piping between such equipment.

The peak demand rate is the maximum rate at which hot water may be drawn in normal service. This may be stated in terms of gallons per minute or gallons per hour of water to be supplied at a temperature of 140°F (60°C). Where water is to be heated instantaneously, such as by a tankless heater, the rated capacity of the heater should be equal to the maximum rate at which hot water is drawn, i.e., the peak demand rate.

A standard method for estimating the peak demand rate for most types of installations has been developed as the result of the experience of heater manufacturers over the years. This method consists of two parts: (1) a base demand rate established for the particular type of installation or occupancy and (2) additional demand rate allowances assigned for certain individual fixtures, groups of fixtures, or occupancy conditions.

For dwellings, a base demand rate of 300 gph (1137 L/h) is recommended currently for tankless systems, and a rate of 180 gph (682 L/h) is recommended for storage systems. These base demand rates may be applied as the peak demand rate per hour in the case of one-family dwellings. No additional demand rate allowances need be added in this case for provision has been made for all the fixtures normally used including an automatic dishwashing machine and a laundry washing machine in a one-family dwelling.

For multiple dwellings, additional demand rate allowances should be assigned as follows: 12 gph (45.4 L/h) for each dwelling unit, 6 gph (22.7 L/h) for each bathroom group of fixtures exceeding the minimum of one group in a dwelling unit, and 3 gph (11.4 L/h) for each automatic domestic dishwashing or laundry washing machine in a dwelling unit. Where laundry washing machines are grouped in a general laundry room in a residential building, an allowance of 30 gph (114 L/h) for each machine should be assigned.

For all other types of building occupancies, a base demand rate of 180 gph (682 L/h) is recommended. Additional demand rate allowances should be assigned as follows:

Bathtubs: 60 gph (227 L/h) each

Laundry trays: 120 gph (454 L/h) for each

Lavatories or shampoo basins: 10 gph (37.9 L/h) for each

Showers: 120 gph (454 L/h) for each in factories, golf clubs, gymnasiums, and schools; 60 gph (227 L/h) for each in business or residence clubs and hospitals; and 12 gph (45.4 L/h) for each in hotels

Sinks: in public restaurants, allow 1½ gph (5.68 L/h) per meal served during the peak period for all sink facilities; in small luncheonettes and soda fountains, allow 1 gph (3.79 L/h) per meal served during the peak period for all sink facilities

Washfountains, multiple type: 260 gph (985 L/h) for each 54-in (1.37 m) circular unit; 180 gph (682 L/h) for each 54-in (1.37 m) semicircular or 36-in (0.91 m) circular unit and 125 gph (474 L/h) for each 36-in (0.91 m) semicircular unit

For tankless heater systems, the duration of the peak draw period need not be considered. This is due to the fact that tankless heaters must be selected of adequate, rated heating capacity to match the peak demand rate regardless of the length of time over which it lasts.

However, for storage tank systems, smaller heaters may be used than in tankless systems, for a storage heater need only provide the heating capacity necessary to maintain proper tank water temperature for the duration of the peak draw period. Over the period of peak draw, water may be drawn from the tank's available supply which should be considered as being 75 percent of the capacity of the tank. For example, if a system has a peak demand rate extending over a 2-h period, 37 percent of the tank's capacity may be drawn per hour. If the period lasts 3 h, 25 percent of the tank's capacity may be drawn per hour. The difference between the peak demand rate per hour and the amount of water which may be drawn per hour from the tank is the heating capacity per hour required for a storage heater.

In large multiple dwellings, the peak demand rate is generally found to be relatively uniform over a maximum period of about 3 h, extending from 6 to 9 A.M. and 5 to 8 P.M. Hence, for such buildings the duration of the peak draw period may be taken as being 3 h.

In smaller buildings occupied for residental use, the period of peak draw is shorter than 3 h. For a one-family dwelling, the period may be assumed to be no more than 20 min in most cases. The period for larger dwellings may be determined by assuming an extra 10 min per each additional dwelling unit in the building until the 3-h period is reached.

For most large buildings having extensive plumbing systems, the period of peak draw may be safely considered to be 3 h. In factory, office, and mercantile buildings, a shorter period may exist in most instances, such as at lunch time and at closing time. The period may not exceed ½ h in some cases, 1 h in others, and 2 h in a few. But, in any case the 3-h period may be assumed when there is no assurance that a shorter period is clearly definable.

Tankless heaters should be selected on the basis of providing rated heating capacity equal to the peak demand rate for the hot water supply system. For example, the tankless heater to select for a modern one family dwelling, with an automatic dishwashing and an automatic laundry washing machine, is one having a rated heating capacity of 300 gph (1137 L/h) or 5 gpm (18.9 L/min).

To select a tankless heater for a 50-family cooperative apartment building, in which 25 dwelling units have an extra bathroom group of fixtures and also dishwashing machines or laundry machines, the procedure is as follows:

	gph	L/h
Base demand rate for multiple dwellings (tankless)	300	1137
Allowance for 50 dwelling units at 12 gph (45.4 L/h)	600	2274
Allowance for 25 extra baths at 6 gph (22.7 L/h)	150	568
Allowance for 25 dishwashing or laundry washing machines at 3 gph (11.4 L/h)	75	284
Total peak demand rate, or required rated heating capacity of tankless heater	1125	4263

Storage heaters should be selected on the basis of providing rated heating capacity equal to the difference between the peak demand rate per hour and the amount of water which may be drawn per hour from the tank. Heater capacities may be determined after appropriate tank sizes have been selected.

For an indirect storage tank heater installation, the minimum storage tank size recommended for a modern one-family dwelling, with an automatic dishwashing and an automatic laundry washing machine, is 66 gal (250 L) capacity. The minimum recommended tank capacities for larger dwellings are as follows: 30 gal (114 L) per dwelling unit for up to 10 units, 25 gal (94.7 L) per unit for up to 25 units, 20 gal (75.8 L) per unit for up to 50 units, and 15 gal (56.8 L) per unit for buildings having more than 50 dwelling units.

To select a storage heater for an indirect storage tank heater installation in a modern one-family dwelling, the heater size should be based upon the use of a 66-gal (250 L) hot water storage tank and a 20-min peak draw period. The recommended peak demand rate for a storage system in a one-family dwelling is 180 gph (682 L/h). So, for the 20-min peak draw period, the peak demand rate is 60 gal (227 L) per 20 min. The amount of hot water available from the tank is 75 percent of its capacity, 66 gal × 0.75 equals 49.5 gal (188 L). The difference between the peak demand rate of 60 gal (227 L) per 20 min and the 49.5 gal (188 L) available from the tank leaves 10.5 gal (39.8 L) of water to be heated in the 20-min period. The heating capacity required on the hourly basis at which heaters are rated is $10.5 \times {}^{60}/_{20}$ or 31.5 gal/h (119 L/h). This is the rated heating capacity per hour of the storage heater recommended for the installation.

Selection of a storage heater for the 50-family cooperative apartment building used as an example in the foregoing may be described. The size of the storage tank recommended for 50 dwelling units is one providing 20 gal (75.8 L) capacity per dwelling unit, or a total capacity of 1000 gal (3790 L). Of this capacity, 75 percent or 750 gal (2842 L) is available for being drawn over a period of 3 h. Hence, the rate of drawing tank water per hour is ${}^{750}/_{3}$ or 250 gph (947 L/h).

For this installation, the total peak demand rate may be established as follows:

	gph	L/h
Base demand rate for multiple dwellings (storage tank)	180	682
Allowance for 50 dwelling units at 12 gph (45.4 L/h)	600	2274
Allowance for 25 extra baths at 6 gph (22.7 L/h)	150	568
Allowance for 25 dishwashing or laundry washing machines at 3 gph (11.4 L/h)	75	284
Total peak demand rate	1005	3808

The rated heating capacity per hour required for the indirect storage heater is the difference between the peak demand rate per hour and the rate at which tank water may be drawn. Thus, the required heating capacity is the difference between 1005 gph (3808 L/h) and 250 gph (947 L/h), or 755 gph (2861 L/h) of rated storage heater capacity.

RETURN CIRCULATION SYSTEMS

Hot water supply piping transmits heat to lower temperature surroundings by convection, radiation, and conduction. This heat is lost by the hot water in the piping. If the water is not reheated and remains stationary in the piping for an extended period, its temperature drops to a degree that the water becomes relatively cold and unsuitable for hot water service. Such water must be drawn from the piping before hot water at normal service temperature is available at outlets. This is objectionable where an excessive length of time is required to purge cold water from the piping before hot water is available, and where an excessive amount of water is wasted in purging cold water from hot water supply systems.

Rarely is there any compelling necessity for circulating hot water continuously through a small system in a one- or two-family dwelling where the length of hot water supply piping is relatively short and pipe sizes are small. In such cases, although hot water may cool off in the piping when no water is drawn, water at normal service temperature is available at outlets within a reasonably short period of time and without excessive drawoff of cool water. For example, if just 1 gal (3.79 L) were drawn, it would be equivalent to purging the contents of 63 ft (19.2 m) of ½-in (12.7-mm) pipe, 36 ft (11.0 m) of ¾-in (19.1-mm) pipe, or 22 ft (6.71 m) of 1-in (25.4-mm) pipe of standard pipe size. Thus, the additional costs involved in installing a return circulation type system and in heating water to compensate for the heat lost by circulation piping are not warranted in most small systems. Of course, where systems supply equipment requiring instantaneous hot water, a return circulation type system may be necessary for satisfactory equipment function.

A return circulation type system definitely is necessary for large extensive hot water supply systems. If they were not designed to circulate hot water continuously, there would be delays in obtaining hot water at normal service temperature, and there would be excessive waste of water of unsuitable temperature. In many instances the length of waiting time could be so great as to cause complaints from building occupants regarding unsatisfactory hot water service where adequate service is paid for in the form of rent. Public water system authorities also object to excessive waste of water because of the general need for conserving the potable water supply, and where buildings are supplied from the public system and are not equipped with water meters, waste may occur without proper charge and cause a corresponding financial loss to the public system.

Consequently, standard code regulations establish a dividing line beyond which return circulation-type hot water supply systems shall be installed. They are required generally in buildings more than four stories in height and in buildings where the developed length of hot water supply piping, measured from the source of hot water supply to the farthest fixture supplied, exceeds 100 ft (30.5 m).

There are three general types of hot water supply-return circulation systems, namely, the upfeed system, the downfeed system, and the combination upfeed and downfeed system. These names are derived from the direction of flow in hot water risers supplying branches to fixtures and are unrelated to the direction of flow in return piping. In each type, return piping is connected to the end or near the end of hot water supply risers so as to circulate water back to the hot water source.

Conventionally, hot water heaters and tanks have been located in cellars and basements for various reasons, such as economy in building design, convenience and economy in locating such equipment in proximity to heating boilers and source of fuel supply, and availability of convenient means for ash disposal. In conventional systems, heaters and tanks are located at the lowest part of the hot water supply system, and under such circumstances, circulation of hot water may be achieved by gravity due to the head induced by the difference in water temperatures in hot water supply and return pipes located above the hot water source. Where systems are extensive or have little effective circulation height, pumps may be used to provide greater circulation.

Inverted design of hot water supply-return circulation systems has been applied advantageously in many tall buildings. In the inverted system, as contrasted with conventional systems, hot water heaters and tanks are located at the highest part of the hot water supply system, and supply and return risers are below the level of the hot water source. Under these circumstances, hot water circulation cannot be achieved by gravity because cooled hot water would settle into the lowest part of the circulation system and remain there. Consequently, in inverted systems circulation of hot water can be achieved only by means of pumps.

The conventional upfeed system is illustrated in Fig. 4-1. In this system, the main hot water supply line is extended from the source of hot water supply and is located in the lowest part of the building. From that location, hot water is supplied to the bottom of all the hot water supply risers. Flow is upward in all the risers supplying branches to fixtures. A hot water return riser is provided for each of the hot water supply risers. The top of the return riser is connected to the supply riser just below the topmost supply branch to fixtures. Return risers extend down to the lowest part of the building where they connect to a main hot water return line through which water circulates back to the source of hot water supply. In this system, air accumulated in the uppermost part of each riser is drawn off when a hot water faucet is opened at a fixture supplied from the top of the supply riser, thereby eliminating air accumulation which might otherwise restrict circulation.

The conventional downfeed system is illustrated in Fig. 4-2. In this system the main hot water supply line is extended from the source of hot water supply to the highest part of the building. From that location,

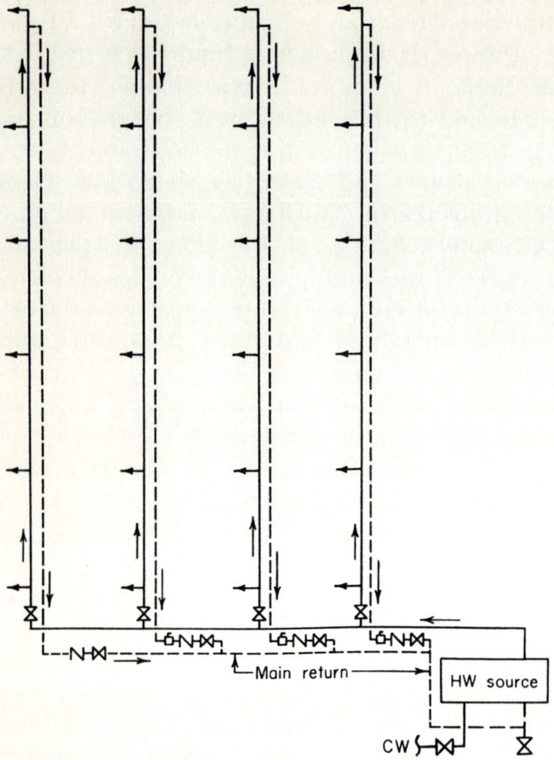

Fig. 4-1 Conventional upfeed system.

hot water is supplied to the tops of all hot water supply risers. Flow is downward in all the risers supplying branches to fixtures. The base of each downfeed riser is connected to a main hot water return line so as to circulate cooled water back to the source of hot water supply. At the top of the main hot water supply line, the highest point in the system, provision should be made to eliminate air so that air pockets do not develop and restrict circulation of hot water. This may be accomplished by connecting a fixture supply branch to the highest point of the system so that air may be drawn off when the hot water faucet is opened at the fixture. Another recommended method is to install an approved type of air relief vent valve at the highest point, with a relief pipe discharging into an approved fixture at which the discharge of air and water from the valve will not be objectionable and where valve leakage will be evident to indicate a need for valve maintenance.

The conventional combination upfeed and downfeed system is illustrated in Fig. 4-3. This system is a combination of the two systems previously discussed. In this case, some hot water supply risers have

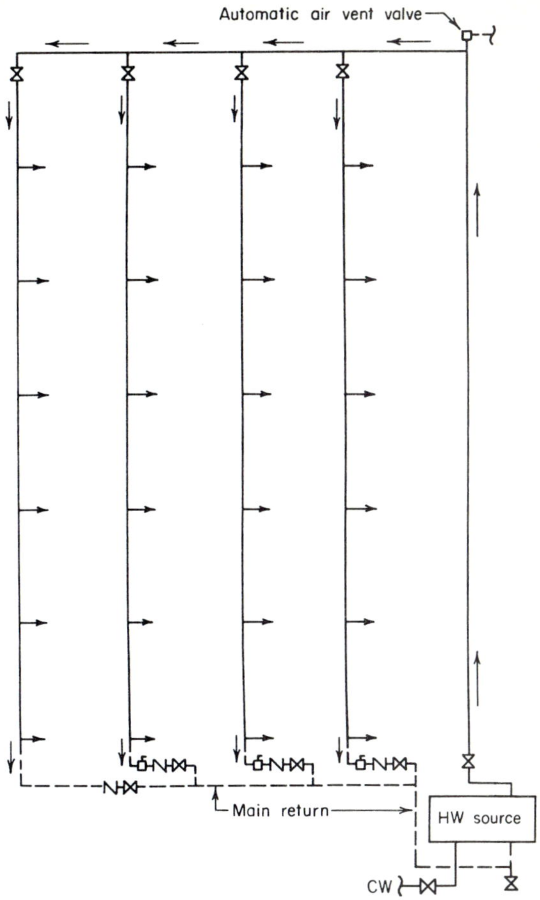

Fig. 4-2 Conventional downfeed system.

upward flow while others have downward flow. Each downfeed riser is supplied from the top of an upfeed riser, and the base of each downfeed riser is connected to a main hot water return line through which water circulates back to the source of hot water supply. Means for preventing air accumulation at high points of the system are necessary. An appropriate method may be selected from those cited for upfeed systems and downfeed systems.

The inverted downfeed system is illustrated in Fig. 4-4. In this system, the main hot water supply line is extended from the source of hot water supply and is located at the highest part of the system. From that location, hot water is supplied to the tops of all the hot water supply risers. Flow is downward in all the risers supplying branches to fixtures. A hot water return riser is provided for each of the hot water supply risers.

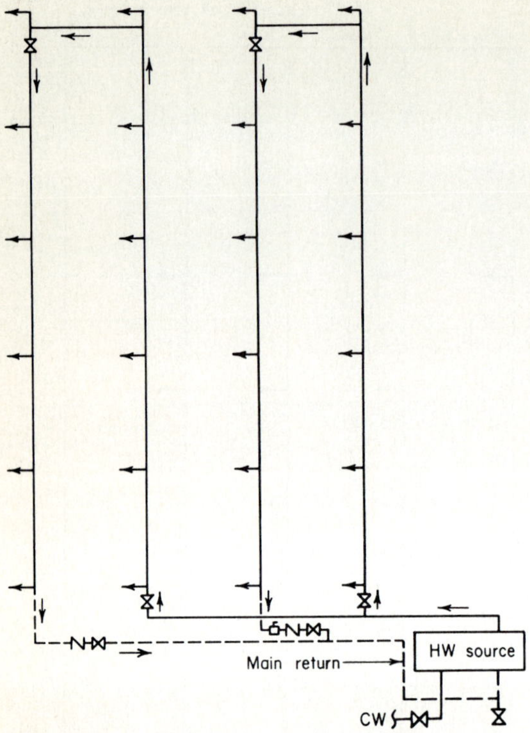

Fig. 4-3 Conventional combination upfeed and downfeed system.

The bottom of the return riser is connected to the supply riser just above the lowest supply branch to fixtures. Return risers extend upward to the highest part of the system where they connect to a main hot water return line through which water is pumped back to the source of hot water supply. Air accumulated in the uppermost part of this system is discharged by means of an automatic air vent valve located at the highest point of the system, thereby eliminating air accumulation which might otherwise tend to restrict circulation, produce noise in the piping, or be objectionable when discharged at a fixture.

Where hot water storage tanks are installed at the highest part of the system, such as is the case in an inverted hot water supply system, it is advisable to provide a tank vacuum breaker, or vacuum relief valve, at the top of the tank so as to permit air to enter the tank whenever a vacuum occurs and thereby prevent collapse of the tank. In such locations, tanks may be subjected to very severe vacuums in the event of cold water supply failure, or when the supply to the tank is shut off and hot water is drawn at a lower floor outlet.

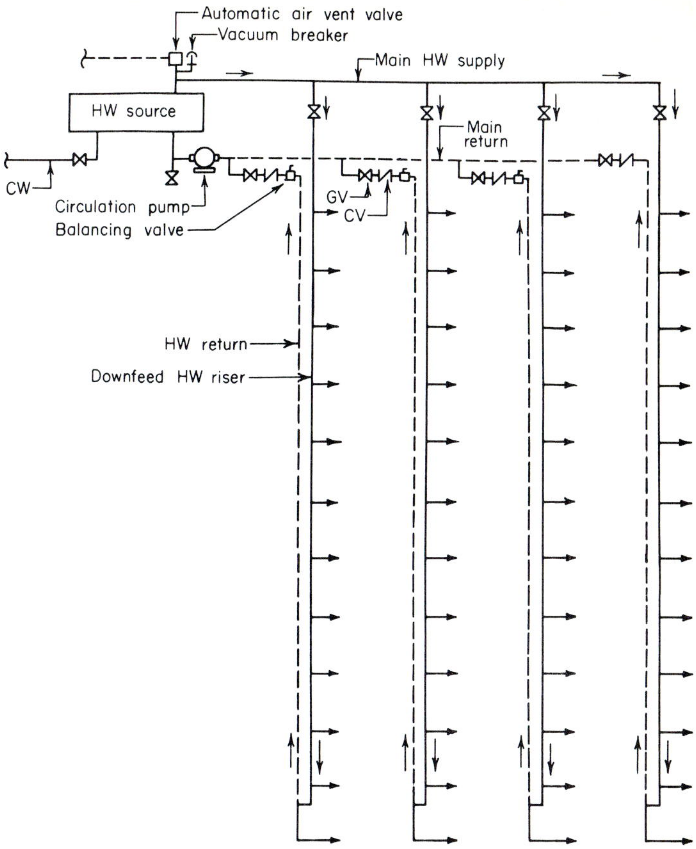

Fig. 4-4 Inverted downfeed system.

The inverted upfeed system is illustrated in Fig. 4-5. In this system, the main hot water supply line is extended from the source of hot water supply to the lowest part of the system. From that location, hot water is supplied to the bottoms of all hot water supply risers. Flow is upward in all the risers supplying branches to fixtures. The top of each upfeed riser is connected to a main hot water return line through which water is pumped back to the source of hot water supply. In this system, air which has accumulated in the uppermost part is automatically discharged by means of an air vent valve located at the highest point of the system, for the same reasons as were given in the case of the inverted downfeed system.

The inverted combination downfeed and upfeed system is illustrated in Fig. 4-6. This system is a combination of the two inverted systems previously discussed. In this case, some hot water supply risers have

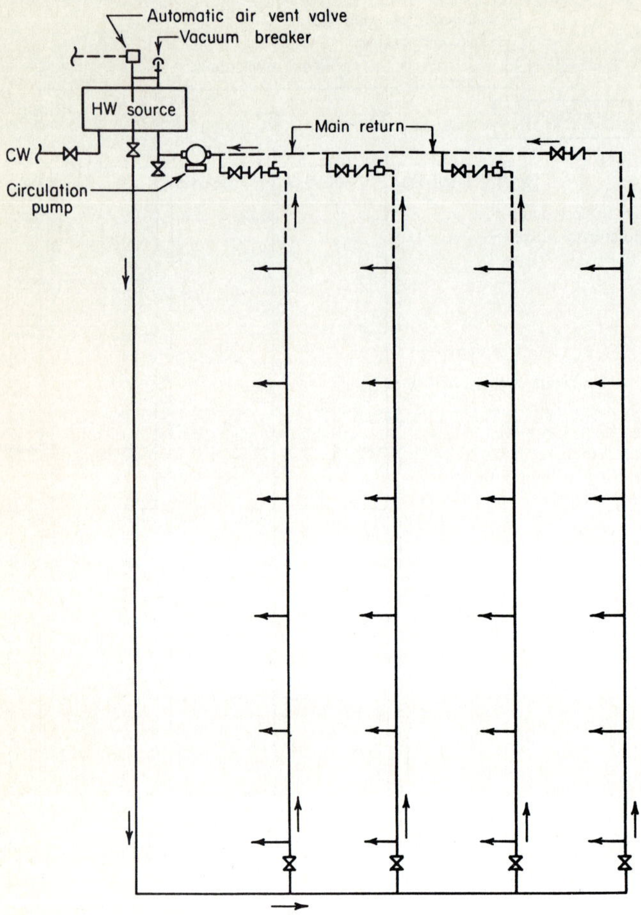

Fig. 4-5 Inverted upfeed system.

downward flow while others have upward flow. Each upfeed riser is supplied from the bottom of a downfeed riser, and the top of the upfeed riser is connected to a main hot water return line through which water is pumped back to the source of hot water supply. Means for eliminating air accumulation in the highest part of the system should be provided for the same reasons as cited for inverted downfeed and inverted upfeed systems.

It is not advisable to connect the main hot water return line to a vertical section of the lower circulation piping between the bottom of a tank and a heater. Water in the main return line generally is hotter than water in the lower circulation piping between the tank and heater, and the hotter return line water tends to rise into the tank while tank

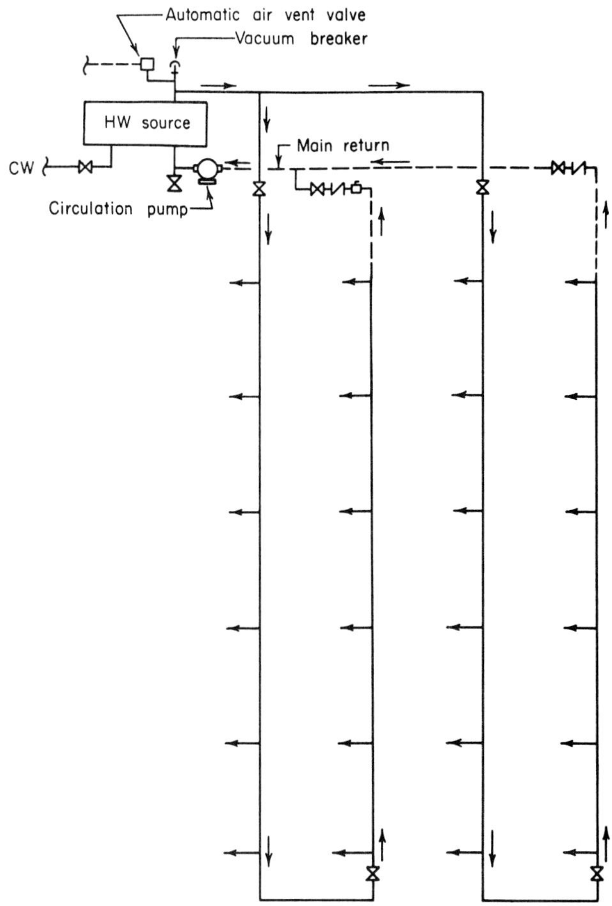

Fig. 4-6 Inverted combination downfeed and upfeed system.

water descends to the heater. These two opposing flows cause circulation to become retarded in both the main return line and in the piping between tank and heater.

Preferably, the main hot water return line of the circulation system should return cooled hot water to the source of hot water supply for reheating by means of a connection to horizontal piping at the heater inlet. Near this connection, a check valve should be provided in the main return line to prevent reverse flow. At the check valve inlet, a gate, ball, or butterfly valve should be installed in the return line and a second gate, ball, or butterfly valve, for blowoff purposes, should be connected to a branch fitting upstream therefrom. By shutting off the gate valve in the main return line and opening the blowoff gate valve, accumulated mud and sediment may be blown out of horizontal sections of the main return line, which should be maintained periodically for satisfactory operation of the system.

Proper valving of all risers is a practical necessity. Gate valves should be provided at the top of all downfeed supply risers and at the bottom of all upfeed supply risers in order to permit such risers to be shut off when necessary. For the same reason, gate valves or plug-type stopcocks should be provided in the horizontal piping at the base of each hot water return riser upstream from its connection to a main hot water return line. In addition, a check valve should be installed adjacent to and upstream from such shutoff valves at the base of each return riser so as to prevent drawing of return line water at fixture outlets.

The force causing circulation in a return circulation type hot water supply system is the difference in head existing in the hot water supply and return lines at the source of hot water supply. In a gravity circulation system, the difference in head is produced by the difference in temperatures, and corresponding densities, of water in the hot water supply and return lines. Water in the return line is cooler and more dense than water in the supply line, and therefore produces a greater hydrostatic head or pressure at the point where they connect to the source of hot water supply. The head induced by the difference in water temperatures in supply and return lines is commonly termed *gravity circulation head,* or at times called *thermal head.* This head varies directly with the heights that the supply and return lines have in common above the heater level.

The recommended design basis for gravity circulation systems is to assume supply and return line temperatures of 140 and 100°F (60 and 37.8°C), respectively. Water density at 100°F (37.8°C) is 61.998 lb/ft^3, and at 140°F (60°C) is 61.386 lb/ft^3. For this 40°F (22.2°C) differential, the density difference is $61.998 - 61.386 = 0.612$ lb/ft^3, and the induced gravity circulation head may be calculated as being $0.612/61.386 = 0.01$ ft (0.003 m) of 140°F (60 °C) water column or head for each foot of effective circulation system height. Thus, if the highest point of the circulation piping were 100 ft (30.5 m) above the hot water tank or heater, the available induced gravity circulation head would be $0.01 \times 100 = 1.00$ ft (0.3 m) of head, based on a 40°F (22.2°C) differential in hot water supply and return line temperatures. This illustrates the fact that the head induced in gravity circulation systems is relatively small, yet sufficiently significant to permit adequate design of such systems where a reasonable amount of effective circulation height exists. Most instances of difficulty in achieving adequate gravity circulation occur in systems which are long in horizontal run and short in effective circulation height, a condition where it is generally advisable to provide positive circulation head by means of a circulation pump.

In pump-equipped return circulation type systems, the pump provides circulation at a rate and head corresponding to the performance characteristics of the pump and the frictional resistance of the circulation piping. Charts showing performance characteristics of various sizes and models of circulation pumps may be obtained from pump manufacturers. From the chart for a given pump, the head available at the pump outlet and the efficiency of the pump may be determined for any rate of discharge within the pump's capacity range. These factors should be considered in selecting pumps suitable for the circulation rates and frictional resistance of systems.

The recommended design basis for systems equipped with circulation pumps is to assume hot water supply and return water temperatures of 140 and 120°F (60 and 49°C), respectively. This 20°F (11°C) differential, when compared with the 40°F (22°C) differential in gravity systems, means that the pump-equipped system is designed to circulate hot water at double the rate of the gravity system and, consequently, provides a higher level of performance in maintaining constancy of hot water supply temperature. This feature should be considered in any comparison of operating and maintenance costs of pump-equipped systems and gravity circulation systems. When comparing the original costs of such systems, the additional amount spent in pump-equipped systems for pumps and their automatic controls should be weighed against the savings resulting from the permissible use of smaller return piping sizes. For example, if a pump-equipped system circulates water at double the rate and at 14.4 times as much circulation head as in a gravity system, the return piping sizes may be calculated as being three-fourths that required for a gravity system.

To balance the temperature of water returning from two or more return risers, so that each returns water to the main hot water return line at the same temperature and at a suitable temperature differential between supply and return piping, is a design objective. However, rarely is it possible to design perfect balance into a system, because exactly correct return pipe diameters cannot be selected from the limited number of standard piping diameters that are available. Hence, means for adjusting circulation rates in risers must be provided in most systems, except those which are relatively small and have a high ratio of circulation height as compared with return piping length. The means for adjusting circulation may be either special balancing valves or plug-type stopcocks.

Figure 4-7 is presented herewith for discussion and reference purposes. In systems having two or more risers, circulation is divided among

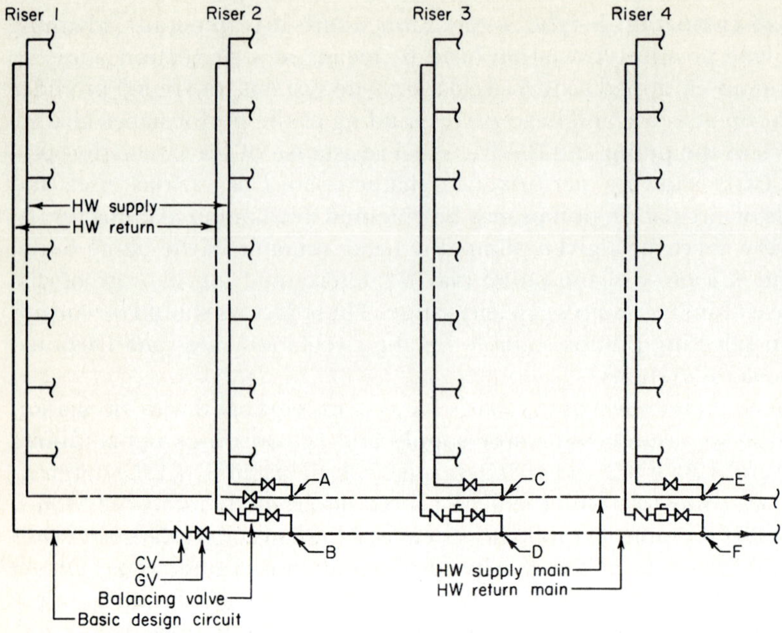

Fig. 4-7 Balancing valves on hot water return risers.

the various risers, which form bypass or parallel circuits through which flow proceeds in accordance with the frictional resistances of the circuits. The riser from which water returns at the lowest temperature is usually at the end of the particular run of return piping having the maximum developed length, and as such, constitutes the basic design circuit of the system of returns.

In the sketch, it may be seen that the distance flow must travel in circulating from point *A*, through riser 1, and back to point *B* is much greater than the distance flow must travel from point *A*, through riser 2, and back to point *B*. In view of the greater distance through riser 1, more frictional resistance occurs in that flow path, and return water temperature will be lower than occurs through riser 2. Similarly, for flow from point *C* to point *D*, more resistance occurs in the flow path through riser 2 than through riser 3; and for flow from point *E* to point *F*, more resistance occurs in the flow path through riser 3 than through riser 4. It may be anticipated that return water temperature at the base of riser 4 will be the highest and at the base of riser 1 will be the lowest.

In order to balance return water temperatures, additional appropriate amounts of frictional resistance must be provided in the returns from risers 2, 3, and 4. This may be done by means of proper adjustment of balancing valves or plug-type stopcocks installed in the horizontal run at the base of those risers. No balancing valve should be provided anywhere in the return piping from riser 1, for this flow path has the greatest frictional resistance in the system; and all other risers in the system must have their circulation rates adjusted so as to achieve return water temperature balance with riser 1.

Before a return circulation system is put into service, it should be adjusted for return line temperature balance under conditions whereby no water is drawn at fixture outlets. The adjustment procedure is simple and may be accurately done using thermometers. First, close all balancing valves so that all circulation goes through riser 1. Then, open the balancing valve on riser 2 to the degree necessary so that it returns water constantly at the same temperature as riser 1. Next, open the balancing valve on riser 3 to the degree necessary so that it returns water constantly at the same temperature as risers 1 and 2. Similarly, adjustment may be made with the balancing valve on riser 4 and with balancing valves on any other risers so that all return water at the same temperature. In this way, relatively perfect return water temperature balance may be achieved.

When a plug-type stopcock is provided as a single device, instead of a gate valve for shutting off the return riser and a balancing valve for adjusting the rate of circulation in the return riser, the stopcock serves a dual purpose. In the event that the stopcock must be shut off during emergencies, it must be properly adjusted when it is open again. Otherwise, the return water temperature balance in the system may be upset sufficiently to cause complaints of inadequate hot water service.

PROCEDURE FOR DETERMINING CIRCULATION RATES AND SIZING RETURN PIPING

The objective of domestic hot water return circulation systems is to maintain hot water temperature within suitable limits in the supply mains and risers. This is done by recirculating hot water through the mains and risers from a storage tank or heater, so that hot water is immediately available at fixtures far removed from the hot water source. To do this economically and accurately, the rates of circulation in, and the sizes of, return piping should be designed in accordance with (1) the heat loss rate of the piping in which circulation occurs, (2) the temperature differential at which the system is to operate, and (3) the pressure or head available for circulation.

A simple, rational method has been developed whereby engineering principles may be applied for determining proper circulation rates and sizes of return piping in any return circulation type domestic hot water supply system to meet particular performance conditions in much the same manner as in hot water heating systems. This method is a complete procedural package, consisting of 11 simple steps that can be applied after hot water supply mains and risers have been sized in accordance with hot water demand requirements of fixtures. The procedure is as follows:

Step 1 Calculate the heat loss rates of the entire hot water supply piping through which hot water is to circulate. This may be done using Table 4-1. It gives heat loss values for various kinds and sizes of piping, bare and insulated, based upon a 140°F (60°C) water temperature and a 70°F (21.1°C) air temperature.

Values for insulated piping assume a covering equivalent to 1 in thick glass fiber. Data compiled in this table are the result of evaluating and converting many items of research information on heat loss from piping gathered bit by bit from scattered reports, charts, manufacturers' literature, and research publications.

Step 2 Determine the heat loss rates of the return mains and risers. As their sizes are unknown at this stage, their heat loss rates cannot be calculated directly. A tentative heat loss rate must be assumed, based upon known facts, experience, and judgment. In this regard, it should be recognized that the length of return piping is approximately the same as that of the supply mains and supply risers in the circulation

Table 4-1
HEAT LOSS RATES FOR HOT WATER SUPPLY PIPING
In Btu/h/linear ft, 140°F Piping to 70°F Air

	Bare			Insulated
Nominal pipe size, in	Galvanized iron or steel pipe, SPS	Brass or copper pipe, SPS or threadless	Copper water tube, type L	All types of piping
½	35	26	19	15
¾	43	32	26	17
1	53	38	32	19
1¼	65	46	39	21
1½	73	53	46	24
2	91	65	58	29
2½	108	75	68	32
3	130	90	81	38
4	163	113	103	46

system. Experience indicates that when final sizes of return piping are established by design, they usually are only about one-half the sizes of comparable sections of supply mains and about three-eights the maximum sizes of supply risers. Table 4-1 may be consulted for the relative heat loss values of the different pipe sizes for such proportions. Based upon these considerations, a reasonable assumption may be made for the heat loss rates of return mains and risers. It is recommended that the assumption be as follows: two-thirds the heat loss rate of the supply piping, when both supply and return piping are insulated, or when both are bare; and four-thirds the heat loss rate of the supply piping, when the supply piping is insulated and the return piping is bare. In this way, heat loss rates for unsized return piping may be tentatively established, and the respective heat loss rates for the various parts of the system may be brought into fairly clear focus.

Step 3 Sum up calculated and assumed heat losses for the entire supply and return piping through which hot water circulates in the system so as to establish its tentative total heat loss rate. Then assign the losses pertaining to individual parts of the system to appropriate main and branch circuits of the circulation piping, in order to establish the proportionate amount of circulation required to compensate for their heat loss loads.

Step 4 Calculate the circulation rates required for the main and branch circuits in accordance with their assigned heat loss loads and with the temperature differential at which the system is to operate. A 20°F (11.1°C) differential is generally recommended for use in designing systems equipped with circulation pumps, while a 40°F (22.2°C) differential is recommended for systems in which circulation is induced by gravity head.

The rate of heat supplied by a circulation rate of 1 gal or 8.3 lb of water per min, in losing 20°F (11.1°C) in temperature, is 166 Btu/min, or 9960 Btu/h. As a matter of convenience, the latter value may be taken as 10,000 Btu/h. The same circulation rate, 1 gpm, in dropping 40°F (22.2°C) supplies about 20,000 Btu/h. These factors may be applied to establish the required circulation rates for all parts of the system. For example, if a pump-equipped system has a total heat loss rate of 61,145 Btu/h, the required pump discharge capacity is equal to 61,145/10,000, or 6.1 gpm; and if the heat loss rate assigned to a given return riser or section of a return main is 8410 Btu/h, the required circulation rate for it is equal to 8410/10,000, or 0.84 gpm. Similar application of the 20,000 Btu/h factor may be made in the case of gravity circulation systems operated at a 40°F (22.2°C) differential. As a result, the circulation rates in gravity systems are just one-half the rates that would be applied if the system were pump-equipped.

Step 5 Determine the pressure or head available for establishing circulation. For a pump-equipped system, this may be determined directly from the performance charts for circulation pumps. Such charts may be obtained from manufacturers. Select the smallest size of pump that has reasonably high efficiency at the required delivery rate. Note particularly the discharge head at which the pump delivers the required rate of flow. For example, if the performance curve for a selected pump shows that the required circulation rate occurs at a discharge pressure of 7.3 ft (2.2 m) of water column or head, this is the head available for circulation.

For a gravity circulation system the available head must be calculated. As a 40° differential is used in designing such systems, the available head may be calculated on the following basis: 0.01 ft (0.003 m) of water column or head multiplied by the vertical height in feet between the elevation of the hot water tank or heater and the elevation of the highest point of the circulation piping. For example, if the highest point of the circulation piping were 109 ft (33. m) above the hot water tank or heater, the available induced gravity head would be calculated as $0.01 \times 109 = 1.09$ ft (0.33 m) of head.

Step 6 Determine which particular run of return piping has the maximum developed length in the system. Measure from the tank or heater to the most extreme point at which the return piping connects to a hot water supply riser. This maximum run of return piping, being the longest in the system, will have the most frictional resistance to flow and require the largest pipe sizes; therefore, it can be called the basic circuit for design purposes. The basic design circuit is illustrated in Fig. 4-8. No balancing valves should be included anywhere in the basic circuit. If they were, someone at sometime certainly would adjust them in a way that would restrict flow, increase the resistance of the basic circuit, and thereby upset the performance of the system. Balancing valves should be provided on all return branches connecting directly to the basic circuit or to branch mains connecting to the basic circuit. Such valves are provided as a means of adjusting and adding resistance to flow in the short return branches so that circulation or flow through them can be reduced to establish temperature balance with the basic circuit.

Step 7 Calculate the pressure drop due to frictional resistance of the

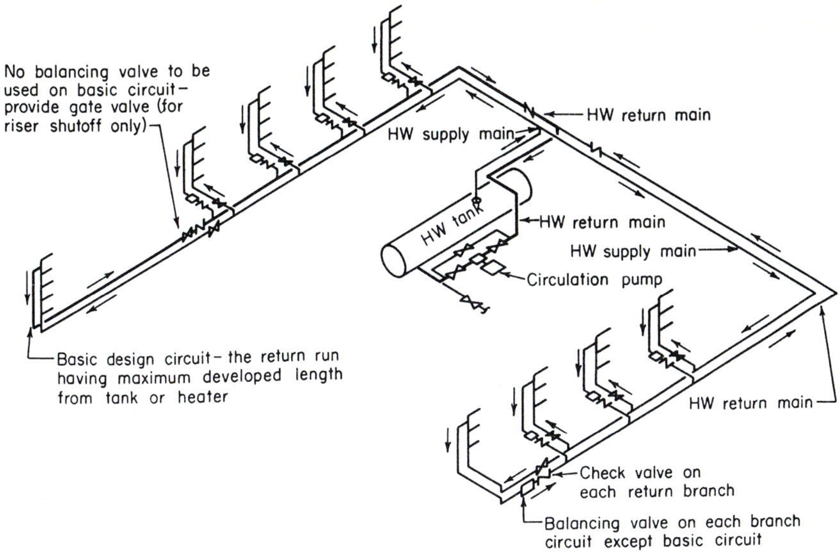

Fig. 4-8 Basic design circuit of hot water recirculation system.

water flowing at required circulation rates in the hot water supply piping extending from the hot water tank or heater, along the supply main, and up the supply riser to the point at which the basic return circuit connects thereto. When this pressure drop is determined (usually found to be so small as to be negligible), it should be deducted from the available head. The difference is the amount of pressure or head which may be dissipated as frictional loss due to flow at the required rates in the basic circuit.

Calculations for pressure drop in the supply piping run may be made directly by applying appropriately one of the following two formulas:

For brass or copper piping:

$$h = 0.000623 q^2 \frac{L}{d^5} \qquad [4\text{-}1]$$

For galvanized iron or steel piping:

$$h = 0.001246 q^2 \frac{L}{d^5} \qquad [4\text{-}2]$$

where h = head or pressure loss due to friction, ft of water column
q = quantity rate of flow, gpm
L = length of pipe, ft
d = diameter of pipe, in

In the above formulas, the length of pipe to be used is not just the developed length of the pipe. Rather it is the sum of the developed length plus the equivalent length calculated for fittings and valves.

Step 8 Determine the maximum permissible uniform pressure drop for the basic circuit. The amount of head that may be dissipated as frictional loss in the basic circuit, as established by step 7, should be divided by the total equivalent length of the circuit. As return pipe sizes are unknown at this stage, the equivalent length for fittings and valves must be assumed tentatively. A reasonable assumption in this case would be 10 percent of the developed length of the basic circuit.

For example, suppose a pump-equipped system is being designed and the performance chart for the pump selected indicates that the pump has a 7.3 ft (2.2 m) discharge head at the required circulation rate; and suppose that the friction loss through the supply piping, as calculated by step 7, is 0.2 ft (0.06 m) of head loss, thereby establishing that 7.1 ft (2.1 m) of head remains for dissipation as frictional loss in the basic circuit. Now, if the basic circuit has a developed length of 645 ft (197 m) and if 10 percent or 65 ft (19.7 m) is assumed as the equivalent length for fittings and valves, the total equivalent length of the circuit may be tentatively established as 710 ft (216.7 m). The maximum uniform pressure drop h/L may then be determined as 7.1 ft (2.17 m) head divided by 710 ft (216.7 m) length, or 0.01 ft (0.003 m) head loss per 1 ft (0.3 m) length.

Similarly, the maximum uniform pressure drop for a system circulated by gravity may be calculated. For example, if the head remaining at the top of the basic circuit was 1.09 ft (0.33 m) and if the total equivalent length of the circuit was 297 ft (90.5 m), the maximum uniform pressure drop h/L would be determined as 1.09 ft head/297 ft length, or 0.00368 ft (0.00112 m) head loss per 1 ft (0.3 m) length.

Step 9 Calculate and tabulate the rates of flow for various pipe sizes, of the type of piping selected for the system, that will produce friction losses corresponding to the maximum permissible uniform pressure drop. Actual internal diameters should be applied. For convenience in making direct flow rate calculations, the two formulas given in step 7 have been rearranged as follows:

For brass or copper piping:

$$q = 40.1 d^{2\ 1/2} \left(\frac{h}{L}\right)^{1/2} \qquad [4\text{-}3]$$

For galvanized iron or steel piping:

$$q = 28.3d^{2\ 1/2}\left(\frac{h}{L}\right)^{1/2} \qquad [4\text{-}4]$$

Using these formulas, a table of flow rates, in gallons per minute, has been developed for various pipe sizes and kinds of piping, based upon a uniform pressure drop h/L of 0.01 ft (0.003 m) head loss per 1 ft (0.3 m) pipe length. Table 4-2 is presented herewith.

Similar tables may be developed for other pressure drops simply by multiplying the flow rates shown in the aforementioned table by an appropriate factor. This may be done because the flow rate is proportional to the square root of the uniform pressure drop h/L. For example, if a table were to be developed to show flow rates corresponding to a uniform pressure drop of 0.00368 ft (0.00112 m) head per ft (0.3 m) length, the appropriate factor would be $(0.00368/0.01)^{1/2} = (0.368)^{1/2} = 0.607$. This factor may be used to multiply all the values given in the table of flow rates based on a pressure drop of 0.01 ft (0.003 m) head per ft (0.3 m) length.

Step 10 Size all parts of the basic circuit. Use the tabulated values of flow rates producing a pressure drop corresponding to the maximum permissible uniform pressure drop for the circuit. The same values may be used for sizing all other parts of the return piping, for such other parts are branch circuits of shorter length than the basic circuit. Thus, the sizes established for branches should be adequate.

Step 11 Now that all return piping sizes have been established, apply these sizes in checking the assumptions and calculations made in steps 2 to 9. Determine the exact heat losses for the system, as there no

Table 4-2
FLOW RATES
Producing a Pressure Drop of 0.01 Ft (0.003 m) Head per 1 Ft (0.3 m) Pipe Length

Nominal pipe size, in	Brass or copper pipe, threadless		Brass or copper pipe, SPS		Copper water tube, type L		Galvanized iron or steel pipe, SPS	
	ID in	Flow, gpm	ID, in	Flow, gpm	ID, in	Flow, gpm	ID, in	Flow, gpm
½	0.710	1.71	0.626	1.24	0.545	.89	0.623	.88
¾	0.920	3.22	0.822	2.44	0.785	2.19	0.824	1.71
1	1.185	6.1	1.06	4.6	1.025	4.26	1.048	3.22
1¼	1.53	11.6	1.37	8.82	1.265	7.22	1.380	6.2
1½	1.77	16.8	1.6	13.1	1.505	11.1	1.61	9.15

longer is any need to rely upon assumptions. Recheck the required circulation rates for all parts of the system, the available circulation head, the total equivalent length of the basic circuit, and the maximum permissible uniform pressure drop. If reasonable assumptions were originally made in this procedure, there should be no reason generally to change sizes any further, except for the few branches that were borderline cases according to the table of flow rates. Such branches may be treated individually in accordance with their loads, length, and the pressure differential between their circuit connections.

STANDARD CODE REGULATIONS

Hot Water Supply System Required Buildings and portions thereof shall be provided with plumbing systems designed to dispose of the sewage from all fixtures and to furnish cold water to every water closet and urinal, and hot and cold water to every sink, laundry tray, automatic laundry washing machine, lavatory, bathtub, and shower required therein, except as otherwise provided herein.

Hot Water for Commercial Dishwashing Machines and Fixtures Commercial dishwashing machines and dishwashing fixtures using hot water shall be provided with water at 140 to 160°F for washing, and at 180 to 190°F for sanitizing.

Pressure Relief Valve An approved pressure relief valve shall be provided on plumbing equipment used for heating or storing hot water. Such valve shall be of adequate relief rating for the equipment served.

Temperature Relief Valve or Energy Shutoff Device An approved temperature relief valve or energy shutoff device shall be provided on equipment used for heating or storing hot water. Temperature relief valves shall be of adequate relief rating for the equipment served. Energy shutoff devices shall be of adequate performance rating for the equipment served.

Combination Pressure-Temperature Relief Valve An approved combination pressure-temperature relief valve of adequate rated relief capacity may be provided in lieu of separate pressure and temperature relief valves.

Location of Relief Valves and Energy Shutoff Devices Pressure relief valves shall be installed in the cold water supply line to the heater or tank served, except that, in areas where scale formation due to water hardness is appreciable, such valve may be installed in an approved location in the hot water supply line from the heater or tank served.

Temperature and combination pressure-temperature relief valves shall be installed so that the temperature-sensitive element is immersed in the hottest water, such as: within the top 6 in of the tank of an underfired hot water storage heater; above the hot water inlet to a tank equipped with a side-arm-type water heater; and above the topmost heating element of an electric water heater.

Relief valves shall be installed so that no check valve or shutoff valve intervenes between a relief valve and the heater or tank served.

Immersion-type energy shutoff devices shall be installed so that the temperature-sensitive element is immersed as required for temperature relief valves. Strap-on-type energy shutoff devices shall be installed so that the temperature-sensitive element is mounted on the tank wall and is responsive to the highest water temperature within the tank.

Relief Valve Outlet Connections No relief outlets of relief valves shall be directly connected to drainage or vent piping. Where pipes from such relief outlets discharge into plumbing fixtures, an air gap shall be provided in conformity with the minimum required air gaps for potable water outlets at plumbing fixtures. No relief outlet or relief pipe shall discharge so as to be a hazard, be a potential cause of damage, or otherwise be a nuisance.

Vacuum Relief Valves Where a hot water storage tank or an indirect water heater is located at an elevation above the fixture outlets in the hot water system, a vacuum relief valve shall be installed on the storage tank or heater.

Location of Pressure Markings on Hot Water Storage Tanks Hot water storage tanks shall be installed so that their pressure markings, showing the maximum allowable water working pressure, are in an accessible location for inspection.

Drain Cocks or Valves for Hot Water Storage Tanks Hot water storage tanks shall be provided with cocks or valves for emptying.

Return Circulation Type System—Where Required Hot water supply systems in other than one- and two-family dwellings shall be of the return circulation type in buildings more than four stories high and in buildings where the developed length of hot water piping, from the source of hot water supply to the farthest fixture supplied, exceeds 100 ft.

Section **5**

Drainage System Design

Drainage systems provide the means for removing from a building the wastes that are collected by the fixtures in a plumbing system. So the design of safe and efficient drainage facilities is critically important for any plumbing system.

This section covers both the external and internal drainage systems needed by any plumbing system. The pipe type and size are discussed, to enable a designer to make the correct choice for the system being designed.

A proven procedure for determining the size of the building drainage system is given. Using this, the designer can develop a safe and economical system for any structure.

5.1 Materials

5.1.1 General. Pipe, tubing, and fittings for drainage systems shall comply with the provisions in Table 5.10.1.

▶ The provisions are those in established standards. Table 5.10.1 provides reference to Federal Specifications or ASA Standards as well as ASTM or Commercial Standards. Each of those standards is applicable to a specific portion of the plumbing system.

5.1.2 Aboveground Piping within Buildings. Soil and waste piping for a drainage system within a building shall be of cast iron, galvanized wrought iron, galvanized open-hearth iron, galvanized steel, lead, brass, or copper pipe, or copper tubing.

5.1.3 Underground Piping within Buildings. Drains within buildings, when underground, shall be of cast-iron soil pipe. For buildings under two stories in height, the pipe may be of service weight. For buildings two stories or more in height, the pipe shall be of extra-heavy weight.

The following materials may be used underground when approved as safe by the administrative authority: galvanized steel or galvanized ferrous alloy, lead, or copper pipe, or copper tubing. Where threaded joints are approved for use underground, they shall be coated and wrapped after installation.

5.1.4 Fittings. Fittings on the drainage system shall conform to the type of piping used. Fittings on screwed pipe shall be of the recessed drainage type.

▶ The piping materials listed in paragraphs 5.1.2 and 5.1.3 are given as guides. **Other kinds can be satisfactory, depending on local conditions, such as local soil characteristics, atmospheric conditions, and specific application to the work involved.**

5.2 Building Sewer

5.2.1 Separate Trenches. The building sewer, when installed in a separate trench from the water-service pipe, shall be cast-iron sewer pipe, vitrified-clay sewer pipe, concrete sewer pipe, bituminized-fiber sewer pipe, or asbestos-cement sewer pipe. Joints shall be watertight and rootproof.

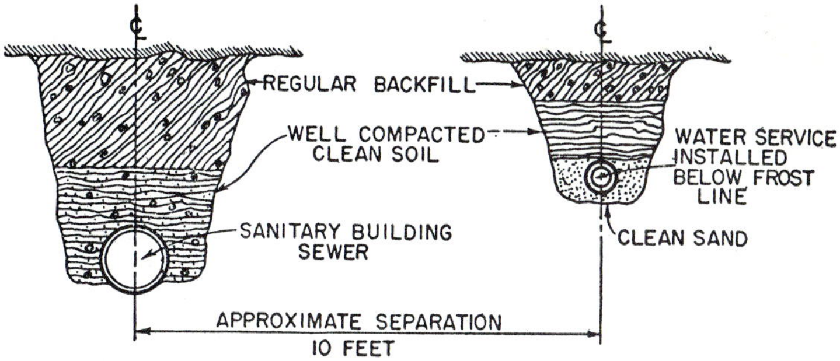

FIG. 5-1

▶ Figure 5-1 illustrates the installation of sanitary sewer and water-service pipe in separate trenches. **Thus, should a break occur in the sewer or water line or a leak develop in poorly made joints, there is small possibility of the potable water being contaminated.**

The building sewer, when installed in a trench *separate* from the water-service pipe, as in Fig. 5-1, may be constructed of any of the following **kinds of pipe:**

(*a*) Asbestos-cement pipe (rubber rings joints)
(*b*) Bituminized-fiber pipe (taper joints)
(*c*) Cast-iron soil pipe (either extra heavy or service weight)
(*d*) Concrete sewer pipe (cement mortar, hot-poured or precast joints)
(*e*) Vitrified-clay pipe (cement mortar, hot-poured or precast joints)

The use of either extra-heavy or service-weight cast-iron soil pipe is satisfactory, but standard-weight pipe may not be installed in buildings over two stories in height. Cast-iron soil pipe has recently been standardized to two weights, extra heavy and service weight.

5.2.2 One Trench. The building sewer, when installed in the same trench with the water-service pipe, shall be constructed of durable materials which are corrosion-resistant and shall be so installed as to remain watertight and be root-proof. The building sewer shall be tested with a 10-foot head of water or equivalent and found to be tight.

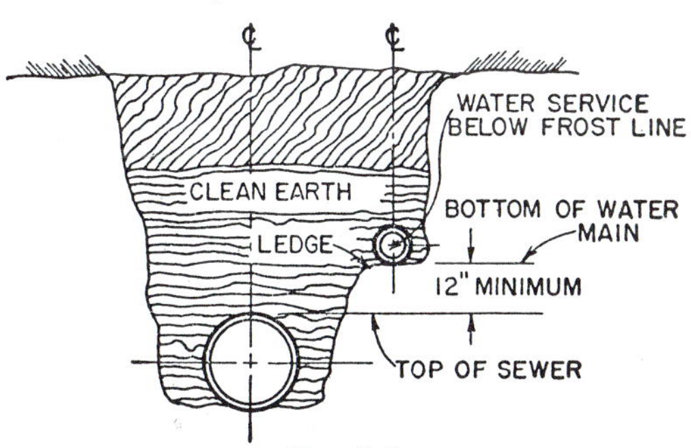

FIG. 5-2

▶ Figure 5-2 shows sewer and water-service pipe in one trench. Because of the additional cost of providing separate trenches for the sewer and water-service pipe, some installations combine both services in the same trench. In order to provide a safeguard for the potable-water main, the water-service main should be placed at least 12 inches above the sewer main and on a side ledge.

The building sewer, when installed in the *same* trench with the water-service pipe, as in Fig. 5-2, shall be constructed of durable materials which are corrosion-resistant and shall be installed so as to remain watertight and root resistant. The following conditions must be met:

(*a*) The water-service pipe must be installed at least 12 inches above the top level of the building sewer pipe.

(*b*) The water-service pipe must be placed on a solid shelf at the side of the trench.

(*c*) The water-service pipe should be in one piece, between the building and the water main.

(*d*) The building sewer may be of any of the following kinds of pipe: extra-heavy cast-iron soil pipe with hot-poured lead joints, hot-poured or precast joints as per paragraph 4.2.6, joints as prescribed in paragraph 4.2.11 for asbestos-cement sewer pipe and as per paragraph 4.2.12 for bituminized-fiber pipe.

(*e*) The building sewer shall be tested after installation with not less than a 10-foot head of water or by means of an equivalent test.

Figure 5-3 shows the excavation necessary under a pipe hub and pipe barrel in firm contact with solid ground.

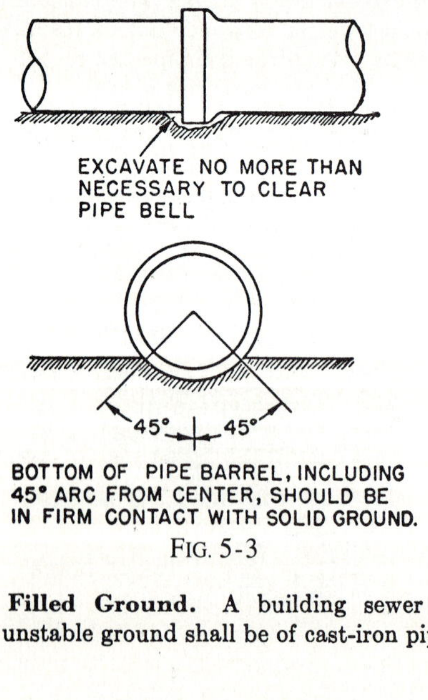

FIG. 5-3

5.2.3 Sewer in Filled Ground. A building sewer or building drain installed in filled or unstable ground shall be of cast-iron pipe, except that nonmetallic drains may be laid upon an approved concrete pad if installed in accordance with paragraph 5.2.1.

▶ **The objective is that piping material which does not have the strength to bridge across unstable ground must be provided with a level support upon which the piping can rest without undue stress.**

Regarding the provision "that nonmetallic drains may be laid upon an approved concrete pad," the necessity for a concrete pad under the piping arises only when the nonmetallic drains are to be installed in filled-in unstable ground under the building.

5.2.4 Sanitary and Storm Sewers. Where separate systems of sanitary drainage and storm drainage are installed in the same property, the sanitary and storm building sewers or drains may be laid side by side in one trench.

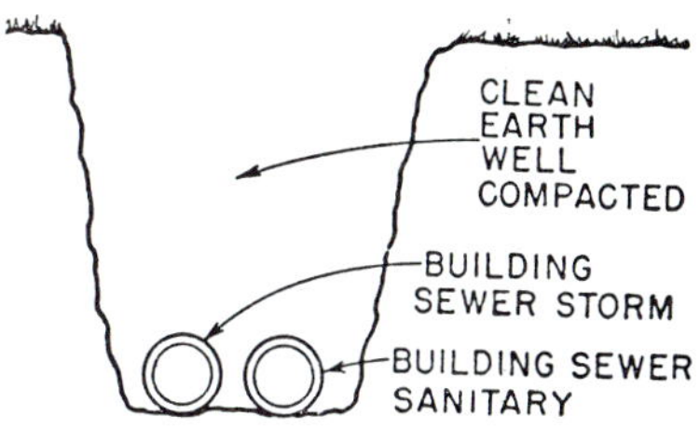

FIG. 5-4

▶ Figure 5-4 illustrates paragraph 5.2.4.

5.2.5 Old House Sewers and Drains. Old house sewers and house drains may be used in connection with new buildings or new plumbing and drainage work only when they are found, on examination and test, to conform in all respects to the requirements governing new house sewers, and the administrative authority shall notify the owner to make the changes necessary to conform to this Code.

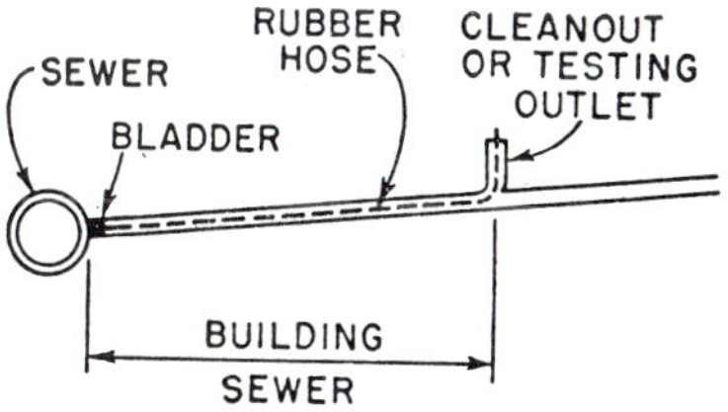

FIG. 5-5

▶ **Before reusing an old sewer it should be tested for flow and tightness. It is possible to test the sewer without excavating. One of the methods is to use a partly filled bladder to which is attached a long rubber hose through which the bladder may be inflated. The bladder is floated to the end of the building sewer. It is then inflated and the sewer pipe filled with water.**

If the pipe and joints are broken, it will not hold a proper level of water. See Fig. 5-5.

5.3 Drainage-piping Installation

5.3.1 Horizontal Drainage Piping. Horizontal drainage piping shall be installed at a uniform slope but at slopes not less than permitted in paragraphs 5.3.2, 5.3.3, and 5.3.4.

5.3.2 Small Piping. Horizontal drainage piping of 3-inch diameter and less shall be installed with a fall of not less than ¼ inch per foot.

5.3.3 Large Piping. Horizontal drainage piping larger than 3-inch diameter shall be installed with a fall of not less than ⅛ inch per foot.

5.3.4 Minimum Velocity. Where conditions do not permit building drains and sewers to be laid with a fall as great as that specified, then a lesser slope may be permitted provided the computed velocity will be not less than 2 feet per second.

► **Generally the slope at which a horizontal drain is installed controls the velocity of the liquid waste flowing in the pipe. A 2-foot-per-second velocity is accepted practice for horizontal waste lines. This velocity will provide the scouring action necessary to maintain a pipe free from fouling.**

Higher velocities, or greater fall per foot, increase the carrying capacity of the building drain. It is a good policy to design for the highest possible velocity, as it will tend to keep the drain pipe clean. When designing fixture branches, however, it should be borne in mind that high velocities in pipes with slopes greater than ¼ inch per foot can cause self-siphonage of the trap seal.

Table A provides a chart of approximate velocities for given slopes and diameters. This table was first computed for BMS 66.

Table A. Approximate Flow Velocity of Sewage for Given Slopes and Pipe Diameters

Diam of pipe (in.)	Flow velocity (fps)			
	1⁄16-in. fall/ft	⅛-in. fall/ft	¼-in. fall/ft	½-in. fall/ft
1¼	....	....	1.61	2.28
1½	....	1.24	1.76	2.45
2	1.02	1.44	*2.03*	2.88
2½	1.14	1.61	2.28	3.23
3	1.24	1.76	2.49	3.53
4	1.44	*2.03*	2.88	4.07
5	1.61	2.28	4.23	4.56
6	1.76	2.49	3.53	5.00
8	*2.03*	2.88	4.07	5.75
10	2.28	3.23	4.56	6.44

House Trap and Fresh-air Inlet. The ASA A40.8 National Plumbing Code and the Report of the Coordinating Committee do not require the installation of a house trap and fresh-air inlet. However, in those cities where house traps are required and a new building is being constructed among old buildings that have house traps, it is not desirable to leave out the house trap and fresh-air inlet in the new building. But where a new community is being planned, the house trap and fresh-air inlet may be omitted.

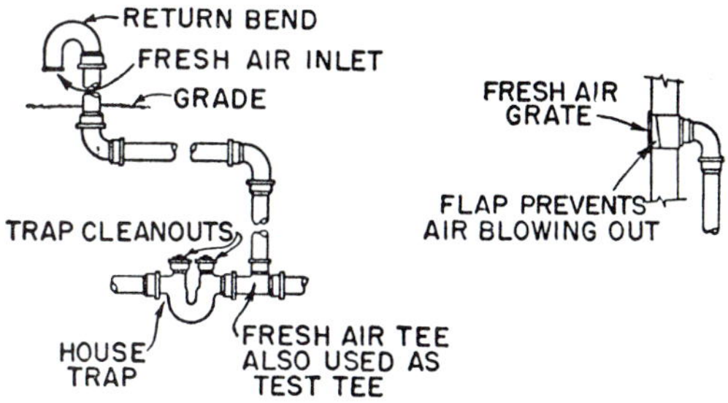

FIG. 5-6

Figure 5-6 illustrates the installation of a house trap and fresh-air inlet. The house trap may be installed in a concrete pit with cover, or the clean-outs may be brought up to the finished floor. The fresh-air inlet may be terminated with a return bend, or if located inside, it may be terminated at the outside front wall with a grate. When a grate is used, install a check or flap which will open readily as an air intake but will not permit a downdraft to blow foul odor into the street.

There are two schools of thought regarding the installation of a building trap in the building drain before connecting it with the building sewer. When a building trap is installed, it is claimed that it will prevent the flow of sewer gases into the plumbing system of the building and thereby avoid using the building system as a means of ventilating the public sewer. It is further claimed that by preventing these sewer gases from entering the building, corrosion of the piping system is reduced.

The fact that a fresh-air inlet is installed when a building trap is provided has proved to be insufficient under certain conditions in relieving the back pressure in the stacks and the building drain. It has been determined by tests that by not installing a building trap the flow and capacity of the building drain are improved.

It is further believed that by permitting the sewer gases to flow freely through the building system the possible concentration of hazardous gases in the street sewer is avoided, thereby preventing possible explosions.

5.4 Fixture Units

5.4.1 Values for Fixtures. Fixture-unit values as given in Table 5.4.2 designate the relative load weight of different kinds of fixtures which shall be

Table 5.4.2 Fixture Units per Fixture or Group

Fixture type	Fixture-unit value as load factors	Minimum size of trap (in.)	
1 bathroom group consisting of water closet, lavatory, and bathtub or shower stall....	Tank water closet 6 Flush-valve water closet 8		
Bathtub[1] (with or without overhead shower)	2		1½
Bathtub[1]............................	3		2
Bidet..............................	3	Nominal	1½
Combination sink and tray..............	3		1½
Combination sink and tray with food-disposal unit..............................	4	Separate traps	1½
Dental unit or cuspidor..................	1		1¼
Dental lavatory........................	1		1¼
Drinking fountain......................	½		1
Dishwasher,[2] domestic..................	2		1½
Floor drains[3]........................	1		2
Kitchen sink, domestic..................	2		1½
Kitchen sink, domestic, with food-disposal unit..............................	3		1½
Lavatory[4]............................	1	Small P.O.	1¼
Lavatory[4]............................	2	Large P.O.	1½
Lavatory, barber, beauty parlor..........	2		1½
Lavatory, surgeon's.....................	2		1½
Laundry tray (1 or 2 compartments)......	2		1½
Shower stall, domestic..................	2		2
Showers (group) per head[2]..............	3		
Sinks:			
Surgeon's..........................	3		1½
Flushing rim (with valve).............	8		3
Service (trap standard)...............	3		3
Service (P trap).....................	2		2
Pot, scullery, etc.[2]..................	4		1½
Urinal, pedestal, siphon jet, blowout......	8	Nominal	3
Urinal, wall lip........................	4		1½
Urinal stall, washout...................	4		2
Urinal trough[2] (each 2-ft section).........	2		1½
Wash sink[2] (circular or multiple), each set of faucets............................	2	Nominal	1½
Water closet:			
Tank-operated.......................	4	Nominal	3
Valve-operated......................	8		3

[1] A shower head over a bathtub does not increase the fixture value.

[2] See paragraphs 5.4.3 and 5.4.4 for method of computing unit values of fixtures not listed in Table 5.4.2 or for rating of devices with intermittent flows.

[3] Size of floor drain shall be determined by the area of surface water to be drained.

[4] Lavatories with 1¼- or 1½-inch trap have the same load value; larger P.O. plugs have greater flow rate.

employed in estimating the total load carried by a soil or waste pipe and shall be used in connection with the tables of sizes for soil, waste, and drain pipes for which the permissible load is given in terms of fixture units.

5.4.2 (See Table 5.4.2.)

5.4.3 Fixtures not listed in Table 5.4.2 shall be estimated in accordance with Table 5.4.3.

Table 5.4.3

Fixture drain or trap size (in.)	Fixture-unit value	Fixture drain or trap size (in.)	Fixture-unit value
1¼ and smaller	1	2½	4
1½	2	3	5
2	3	4	6

▶ It is difficult to provide a rating for all plumbing fixtures as shown on Table 5.4.2 because of the larger number of fixtures made for special purposes, such as printers' sinks, laboratory sinks, and special hospital fixtures. Special-purpose fixtures may be rated by the size of the trap recommended by the manufacturer. It is hoped that eventually all manufacturers of fixtures and appliances will indicate the fixture-unit value on their products. Such values for each fixture would greatly assist in accurately designing a plumbing system, especially for large buildings. In the meantime, Table 5.4.3 provides an approximate rating of those fixtures not rated in Table 5.4.2.

5.4.4 Values for Continuous Flow. For a continuous or semicontinuous flow into a drainage system, such as from a pump, pump ejector, air-conditioning equipment, or similar device, two fixture units shall be allowed for each gallon per minute of flow.

▶ Figure 5-7 illustrates sizing for an intermittent flow from a sump pump. The building has five soil stacks. Stacks 1, 2, and 5 contain public-type fixtures. Stacks 3 and 4 contain private-type fixtures. The total fixture units for each stack are given.

Each stack is sized according to Table 5.5.3. Horizontal branches from the base of a stack to the building drain are the same size as the building drain (Table 5.5.2), and the building drain is then sized according to its fall per foot.

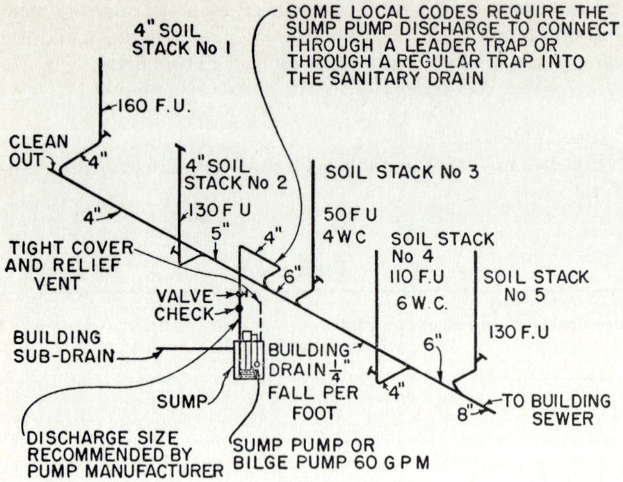

FIG. 5-7

The discharge from the sump is converted into fixture-unit values by multiplying the number of gallons flow by 2. It is recommended, however, where the discharge from the sump is connected to the building drain, that there be no connections from the gravity system for at least 10 feet downflow from the sump connection. If this is impractical, the building drain should be increased one pipe size. This prevents a heavy load concentration which could affect the fixtures on the lower floor where the stack is connected close to the sump discharge.

It would be preferable to connect large flows from the sump close to the point where the building drain leaves the building. Where the discharge is from pneumatic ejectors, or where heavy and abrupt discharges are to be expected, the discharge should be carried separately to the building sewer just outside the building wall.

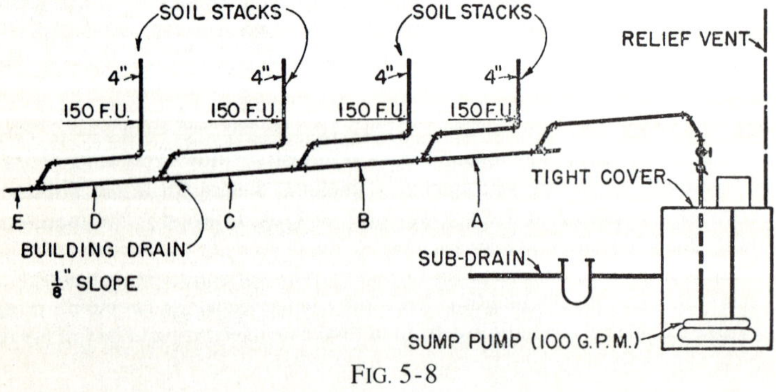

FIG. 5-8

Figure 5-8 illustrates another example of continuous-flow rate. The number of fixture units equivalent to a sump pump discharging 100 gallons per minute is 200 fixture units. According to Table 5.5.2, the drain may be sized as follows:

Sizing of Drain
Example: Fig. 5-8

Slope: 1/8 in./ft			Slope: 1/4 in./ft		
Location	Number of fixture units	Diameter of drain (in.)	Location	Number of fixture units	Diameter of drain (in.)
Point A	200	5	Point A	200	4
Point B	350	5	Point B	350	5
Point C	500	6	Point C	500	6
Point D	650	6	Point D	650	6
Point E	800	8	Point E	800	6

It is desirable that the slope of the building drain be maintained uniformly throughout its length. If this is impractical, select a pipe size which provides for velocities of not less than 2 feet per second. (See Table A.)

Horizontal piping from a branch to a group of fixtures on the same level as the building drain should be the same as for the building drain. (See paragraph 5.5.2.)

A 3-inch building drain is large enough to serve two water closets, plus whatever additional small fixtures are desired, provided the total number of fixture units does not exceed the maximum permitted in Table 5.5.2. Blowout-type urinals are classified as water closets in computing the size of the building drain. The reason for this is that a blowout urinal will discharge as much water as a regular water closet in the same time.

5.5 Determination of Sizes for the Drainage System

5.5.1 Maximum Fixture-unit Load. The maximum number of fixture units that may be connected to a given size of building sewer, building drain, horizontal branch, or vertical soil or waste stack is given in Tables 5.5.2 and 5.5.3.

5.5.2 (See Table 5.5.2.)
5.5.3 (See Table 5.5.3.)

▶ The following is an explanation of the theory which resulted in the establishment of Tables 5.5.2 and 5.5.3:

Developments inside a drainage and venting system are complicated. Individual fixtures have different rates of discharge due either (1) to the average rate of flow each time the fixture is used or (2) to the length of time of each discharge.

Table 5.5.2 Building Drains and Sewers

Diameter of pipe (in.)	Maximum number of fixture units that may be connected to any portion of the building drain or the building sewer[1] — Fall per foot			
	1/16 in.	1/8 in.	1/4 in.	1/2 in.
2	.....	.....	21	26
2½	.....	.....	24	31
3	.....	20[2]	27[2]	36[2]
4	.....	180	216	250
5	.....	390	480	575
6	.....	700	840	1,000
8	1,400	1,600	1,920	2,300
10	2,500	2,900	3,500	4,200
12	3,900	4,600	5,600	6,700
15	7,000	8,300	10,000	12,000

[1] Includes branches of the building drain.
[2] Not over two water closets.

Table 5.5.3 Horizontal Fixture Branches and Stacks

Diameter of pipe (in.)	Maximum number of fixture units that may be connected to:			
	Any horizontal[1] fixture branch	One stack of 3 stories in height or 3 intervals	More than 3 stories in height	
			Total for stack	Total at one story or branch interval
1¼	1	2	2	1
1½	3	4	8	2
2	6	10	24	6
2½	12	20	42	9
3	20[2]	30[3]	60[3]	16[2]
4	160	240	500	90
5	360	540	1,100	200
6	620	960	1,900	350
8	1,400	2,200	3,600	600
10	2,500	3,800	5,600	1,000
12	3,900	6,000	8,400	1,500
15	7,000			

[1] Does not include branches of the building drain.
[2] Not over two water closets.
[3] Not over six water closets.

If it were normal practice to operate all fixtures at exactly the same moment, it would be simple to estimate what the total discharge flow would be at a given moment, but in a large building with many plumbing fixtures, all the fixtures never would be started and stopped exactly at the same moment under ordinary working conditions.

The theory that the fall of liquids within a stack continues to accelerate like a free-falling object is erroneous. As soon as the water from a horizontal drain enters the vertical stack, the force of gravity accelerates it downward; but within a short distance, usually within a one- or two-story drop, the friction against the interior walls of the pipe becomes equal to the force of gravity, thereby equalizing the rate of fall. From that point of equilibrium, the rate of fall down the stack continues unchanged until it reaches the base of the stack, provided, of course, that no lower horizontal branches also commence discharging, colliding, and interfering with the fall.

Therefore, the pipe sizes in Tables 5.5.2 and 5.5.3 are a reconcilement of the following considerations:

(*a*) Experience and good practice have established that the "normal fullness" of piping in first-class condition is one-quarter full.

(*b*) The remaining capacity is the margin of safety for unusual peak demand.

(*c*) Use the smallest size pipe that would carry away waste water from individual fixtures as rapidly as is consistent with good practice to avoid clogging.

(*d*) Avoid causing excessive pressures at points where the fixture drains connect to the stack. Pressures either negative or positive could reduce the trap seal to a dangerous level or blow sewer gas into the room.

5.5.4 Minimum Size of Soil and Waste Stacks. No soil or waste stack shall be smaller than the largest horizontal branch connected thereto, except that a 4 by 3 water-closet connection shall not be considered as a reduction in pipe size.

▶ Figure 5-9 illustrates a one-story house containing four dwelling units with the building drain underground.

In section *a* the water closet is stack-vented; the bathtub is wet-vented through the lavatory; the sink and dishwasher have a common vent. See paragraph 12.10.2 for sizing when one plumbing fixture connects below the other. If both plumbing fixtures connect at the same level, the vertical waste is sized according to Table 5.4.2 for the total number of fixture units; in this case, there are four fixture units, which require a 1½-inch waste.

In section *b* the water closet is stack-vented; the bathtub is wet-vented through the lavatory waste; the sink and food-waste grinder have a common vent. The vertical waste below the sink connection is sized 2 inches. If both fixtures are connected at the same level, the vertical waste is 2 inches, because the total number of fixture units is five.

All fixtures in *c* are stack-vented except the sink, which is individually vented or back-vented. The back vent is 1¼ inches because the sink represents only two fixture units. The stack here must be 3 inches up to the highest fixture connection.

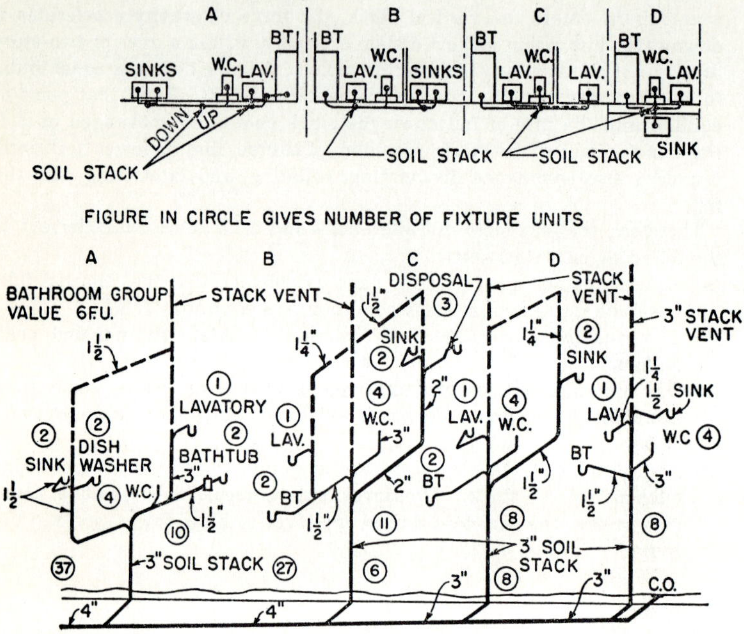

FIG. 5-9

In section *d* all fixtures are stack-vented, and regardless of climate the soil stack *must* be extended full size 3 inches in diameter through the roof as a stack vent.

The building drain is sized according to Table 5.5.2, and cleanouts are provided according to paragraph 5.4.1. Branches for connecting plumbing fixtures are vented at the distances provided in the Code. The slope at which a branch is roughed should not exceed one pipe diameter.

There are other ways of wet venting a fixture which cannot be provided with a continuous or back vent. The principle of wet venting may be applied to a floor drain, a dental unit, or any other fixture which must be installed in the middle of a room and cannot be individually vented.

The advantage of wet venting is that it ensures a clear vent. There is no assurance that a back vent installed horizontally will remain open and effective, nor is there any way of knowing when it becomes clogged.

5.5.5 Minimum Size of Stack Vent or Vent Stack. Any structure on which a building drain is installed shall have at least one stack vent or vent stack carried full size through the roof not less than 3 inches in diameter or the size of the building drain, whichever is the lesser.

▶ By reference to Fig. 5-9 it will be noted that stack *d* is shown full size through the roof but other stack vents may be reduced.

5.5.6 Future Fixtures. When provision is made for the future installation of fixtures, those provided for shall be considered in determining the required sizes of drain pipes. Construction to provide for such future installation shall be terminated with a plugged fitting or fittings at the stack so as to form no dead end.

5.5.7 Underground-drainage Piping. No portion of the drainage system installed underground or below a basement or cellar shall be less than 2 inches in diameter.

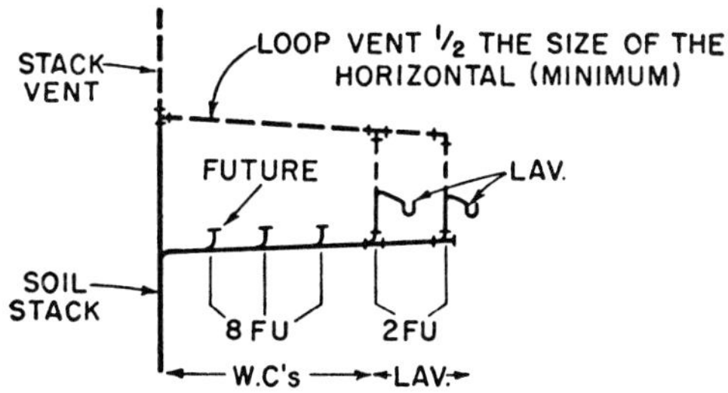

FIG. 5-10

▶ Figure 5-10 illustrates a toilet room with two water closets and two lavatories to be installed and roughing for a third water closet to be installed at some future time. The branch must therefore be sized as though all three water closets were to be installed initially.

The total number of fixtures now installed are two public water closets and two public lavatories with a fixture rating of 8 + 8 + 2 + 2 = 20 fixture units. Table 11.5.3, column 2, any horizontal fixture branch, shows that a 3-inch diameter pipe is satisfactory. However, by roughing a future closet on this branch, the total number of fixture units that must be figured are 8 + 8 + 8 + 2 + 2 = 28 fixture units; therefore the diameter of the branch must be 4 inches.

There is also a difference in sizing the loop vent. With a 3-inch horizontal branch a 1½-inch loop vent is adequate. With a 4-inch horizontal waste, a 2-inch-diameter loop would be the minimum permitted.

5.6 Offsets on Drainage Piping

5.6.1 Offsets of 45 Degrees or Less. An offset in a vertical stack, with a change of direction of 45 degrees or less from the vertical, may be sized as a straight vertical stack. In case a horizontal branch connects to the stack within 2 feet above or below the offset, a relief vent shall be installed in accordance with paragraph 12.18.3.

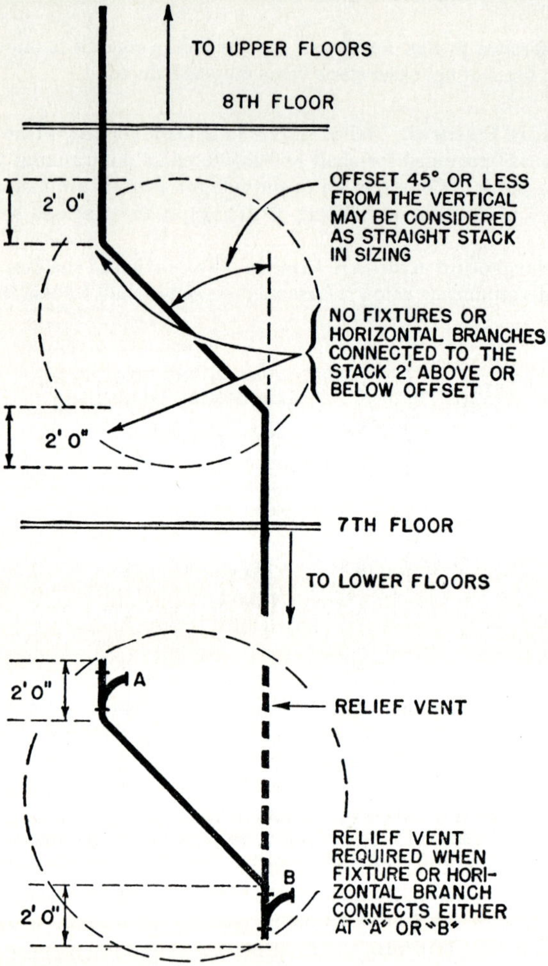

FIG. 5-11

Figure 5-11 shows that the relief vent may be installed either as a vertical continuation of the lower section of the stack or as a side vent connected to the lower section between the offset and the next lower fixture or horizontal branch. The diameter of the relief vent must be no less than the diameter of the main vent or the soil or waste stack, whichever is the smaller.

Offsets of 45 degrees or less do not seriously affect the flow in the stack. Table 5.5.3 provides for loading of stacks not more than ½-full at any point. This partial loading is a safety factor. When a horizontal load is imposed at this point, the relief vent is necessary in order to prevent excess pressures from affecting the operation of the fixtures connected to the branch.

5.6.2 Waste Stacks Serving Kitchen Sinks. In a one- or two-family dwelling only in which the waste stack or vent receives the discharge of a kitchen-type sink and also serves as a vent for fixtures connected to the horizontal portion of the branch served by the waste stack, the minimum size of the waste stack up to the highest sink branch connection shall be 2 inches in diameter. Above that point the size of the stack shall be governed by the total number of fixture units vented by the stack.

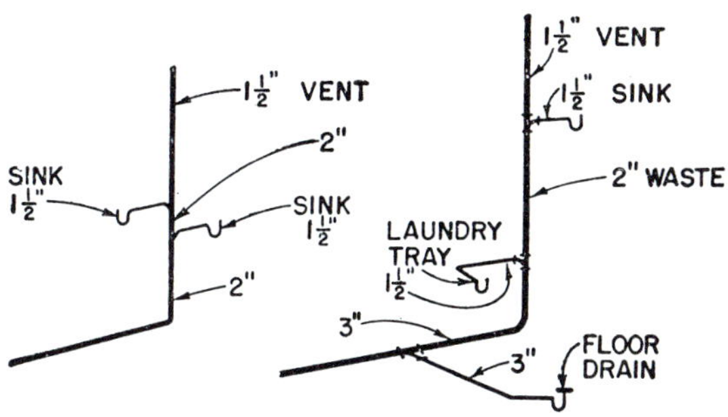

FIG. 5-12

5.6.3 Above Highest Branch. An offset above the highest horizontal branch is an offset in the stack vent and shall be considered only as it affects the developed length of the vent.

▶ Figure 5-13 shows that offsets occurring above the highest fixture or horizontal branch are part of the stack vent and do not affect the capacity of the stack. The vent at this point is of sufficient diameter to provide for air circulation.

5.6.4 Below Lowest Branch. In the case of an offset in a soil or waste stack below the lowest horizontal branch, no change in diameter of the stack because of the offset shall be required if it is made at an angle not greater than 45 degrees. If such an offset is made at an angle greater than 45 degrees, the required diameter of the offset and the stack below it shall be determined as for a building drain. (See Table 5.5.2.)

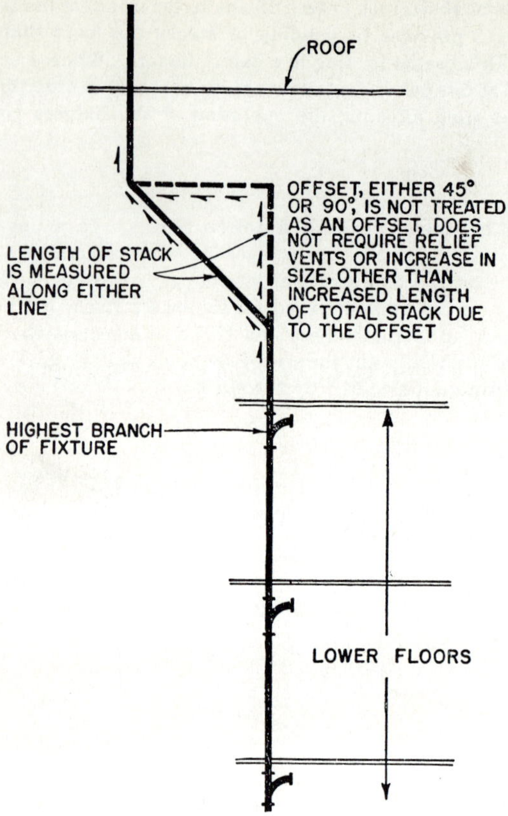

FIG. 5-13

▶ **Figure 5-14 illustrates an offset below the lowest horizontal branch. A 45-degree offset will not affect the size of the pipe, regardless of where such an offset occurs in the soil or waste stack. But if the offset is in a horizontal position (see definition of horizontal pipe), it is necessary to increase the diameter of the offset pipe.**

This is done to avoid developing pressure or vacuum which will disturb the trap seal of the lower fixture. The threat of such a condition becomes more serious as the load in the stack is increased. The most critical area is at the base of the stack or at the lowest floor in a building, especially when an offset or change of direction occurs immediately below a plumbing fixture. A plumbing fixture should not be installed nearer than 2 feet from the offset.

5.6.5 Offsets of More Than 45 Degrees. A stack with an offset of more than 45 degrees from the vertical shall be sized as follows:

The portion of the stack above the offset shall be sized as for a regular stack based on the total number of fixture units above the offset.

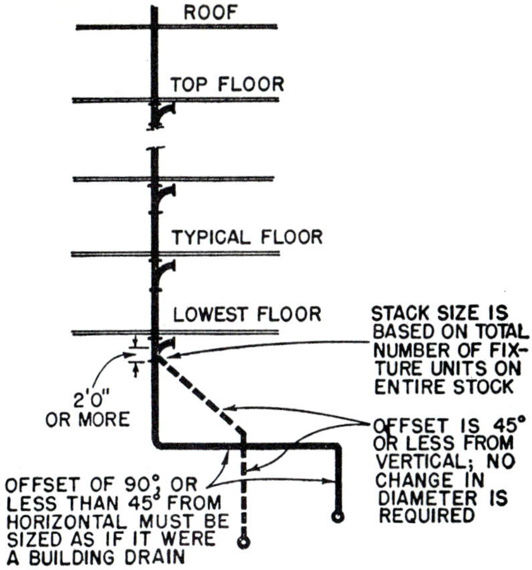

FIG. 5-14

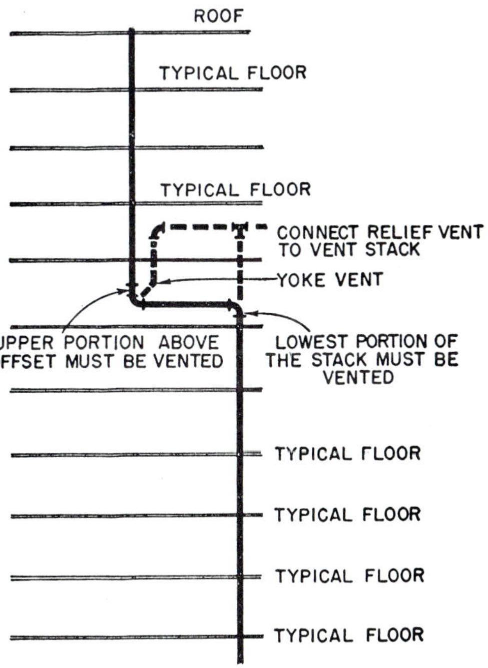

FIG. 5-15

The upper portion of the stack above the offset shall be sized as for building drain. (See Table 5.5.2, column 5.)

The portion of the stack below the offset shall be sized as for the offset or based on the total number of fixture units on the entire stack, whichever is the larger. (See Table 5.5.3, column 4.)

A relief vent for the offset shall be installed as provided in the Code, and in no case shall the horizontal branch connect to the stack within 2 feet above or below the offset.

▶ Figure 5-15 illustrates an offset of more than 45 degrees; in this case the offset is 90 degrees.

The upper portion of the offset must be provided with a relief vent in order to avoid back pressure affecting the fixtures located near the offset. The lower portion of the stack must be provided with a relief vent in order to safeguard the fixtures on the downstream portion of the stack.

In order clearly to illustrate the requirements of paragraph 5.6.5, Fig. 5-16 shows the sizing of a stack in a 12-story building where there is one offset between the fifth and sixth floors and another offset below the street floor.

Sizing is computed as follows:

Step 1: Compute the fixture units connected to the entire stack. In this case, assume there are 1,200 fixture units connected to the stack from the street floor through the top floor.

Step 2: Size the portion of the stack above the fifth-floor offset. There are 400 fixture units from the top floor down through the sixth floor. According to Table 5.5.3, column 4, 400 fixture units require a 4-inch stack.

Step 3: Size the offset on the fifth floor. An offset is sized like a building drain. According to Table 5.5.2, column 4, 400 fixture units require a 5-inch offset.

Step 4: Size the lower portion of the stack from the fifth floor down through the street floor. The lower portion of the stack must be large enough to serve all the fixture units connected to it, from the top floor down, in this case, 1,200 fixture units. According to Table 5.5.3, 1,200 fixture units require a 6-inch stack.

Step 5: Size the offset below the street floor the same as for a building drain. The lower offset also has to be large enough to serve all fixture units from the top floor down, in this case, 1,200 fixture units. According to Table 5.5.2, 1,200 fixture units require an 8-inch offset. This 8-inch line is run full size to the building drain.

The fixtures on the sixth floor should be connected to the stack at least 2 feet above the offset. If this is not possible, then connect them separately to the stack at least 2 feet below the offset, and if this is not possible either, run the fixture drain down to the fifth or fourth floor and connect to the stack there.

The offset on the fifth floor should be provided with a relief vent. Sizing the offset larger than the stack and providing a relief vent will prevent pressures from building up at the point of offset and possibly siphoning or blowing nearby trap seals.

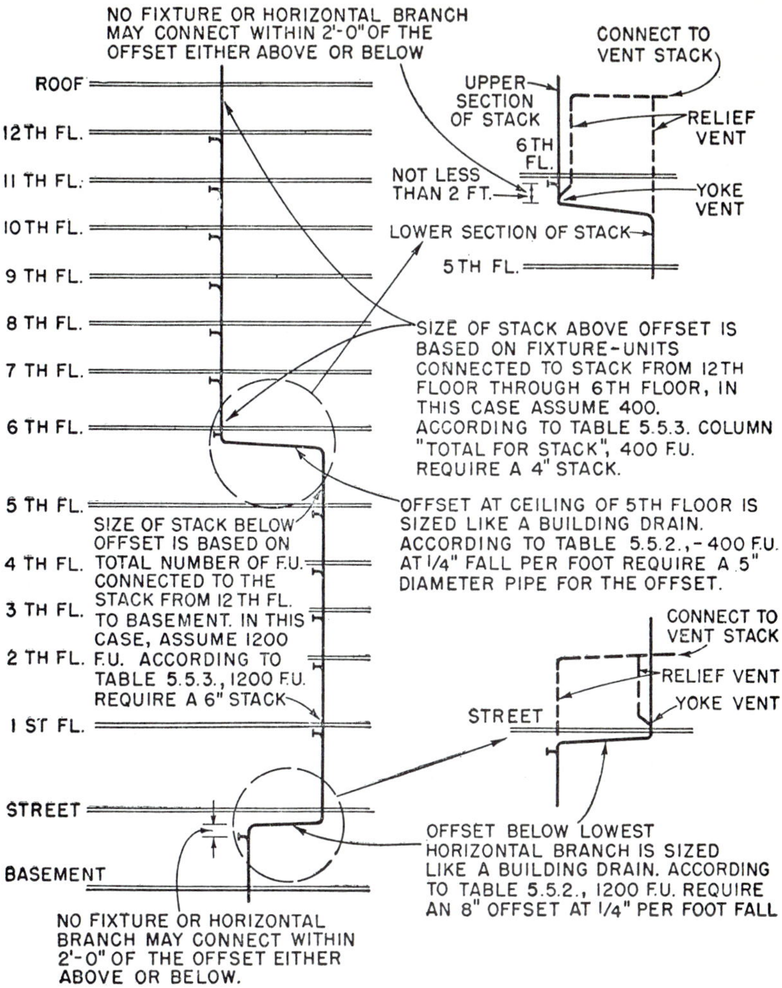

FIG. 5-16

The critical points of a soil stack are at the base and where an offset occurs. Provide ample relieving vents at these points. Each branch should be properly sized and vented to prevent an unbalanced condition within the branch.

5.7 Sumps and Ejectors

5.7.1 Building Drains below Sewer. Building drains which cannot be discharged to the sewer by gravity flow shall be discharged into a tightly covered and vented sump, from which the liquid shall be lifted and discharged into the building gravity drainage system by automatic pumping equipment or by any equally efficient method approved by the administrative authority.

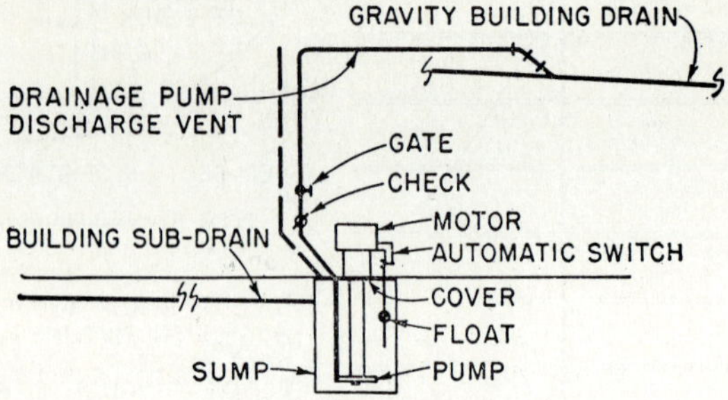

FIG. 5-17

▶ **Sewage below the level of the sewer is generally handled through automatically controlled sewage pumps or air ejectors.**

Figure 5-17 illustrates a simple installation for handling drainage which cannot be discharged by gravity into the public sewer.

When a limited number of fixtures are installed below sewer level, a single pump will be sufficient to handle the work; but when a large number of fixtures are involved, a duplex system should be used, so that in the event of a breakdown of one pump another would remain in operation. When a duplex unit is used, it is good practice to alternate the operation of the pumps automatically.

5.7.2 Storage Period. The storage of drainage in a sump or ejector shall not exceed a period of 12 hours.

▶ **The storage period depends upon the number of fixtures installed in the sump or ejector system. There are times, for instance over a week end, when in industrial or commercial buildings the system need not operate regularly.**

5.7.3 Design. Sump and pumping equipment shall be so designed as to discharge all contents accumulated in the sump during the cycle of emptying operation.

▶ **If an automatic system is designed, it is generally arranged so that it will be in continuous operation until the contents of the tank are pumped out.**

5.7.4 Venting. The system of drainage piping below the sewer level shall be installed and vented in a manner similar to that of the gravity system.

▶ **The tables applying to the gravity system also apply to a subsoil system.**

5.7.5 Duplex Equipment. Sumps receiving the discharge of more than six water closets shall be provided with duplex pumping equipment.

▶ **Six water closets have been determined as the minimum to be installed with a single pump. There are exceptions to this rule, which must be judged by occupancy and availability of other facilities.**

5.7.6 Vent Sizes. Building sump vents shall be sized in accordance with Table 12.21.5 but shall in no case be sized less than 1½ inches.

5.7.7 Separate Vents. Vents from pneumatic ejectors or similar equipment shall be carried separately to the open air as a vent terminal.

5.7.8 Connections. No direct connection of a steam exhaust, blowoff, or drip pipe shall be made with the building drainage system. Waste water, when discharged into the building drainage system, shall be at a temperature not higher than 140°F. When higher temperature exists, proper cooling methods shall be provided.

▶ **Pneumatic ejectors have a tendency to discharge quickly, differing from the even flow of the centrifugal pumping equipment. It is therefore recommended that, where compressed-air ejectors are used, the venting system be separate from the gravity system.**

Paragraph 5.7.8 protects the building piping as well as the building and public sewer. **Extremely hot water or steam discharges may damage the piping. Hot-water discharge at more than 140°F or steam-discharge should first be collected into a sump pump, where it may cool off before flowing into the sewer.**

5.8 Floor Drains

5.8.1 Accessibility. Floor drains shall connect into a trap so constructed that it can be readily cleaned and of a size to serve efficiently the purpose for which it is intended. The drain inlet shall be so located that it is, at all times, in full view.

5.8.2 Connection. Floor drains subject to backflow shall not be directly connected to the drainage system.

5.8.3 Provision for Evaporation. Floor-drain trap seals subject to evaporation shall be of the deep-seal type or shall be fed from an approved plumbing fixture or by means of an approved automatic priming device designed and approved for that purpose.

▶ **The trap shall be either accessible from the floor-drain inlet or by a** separate cleanout within the drain. Figure 5-18 illustrates several types **of drains which meet these conditions.**

(*a*) **Typical drain with integral trap, that may be cleaned through removable strainer at floor level.**

(*b*) **Floor drain with combination cleanout and backwater valve, where there is a possibility of backflow.**

(*c*) **Drain with combined cleanout, backwater valve and secondary bucket.**

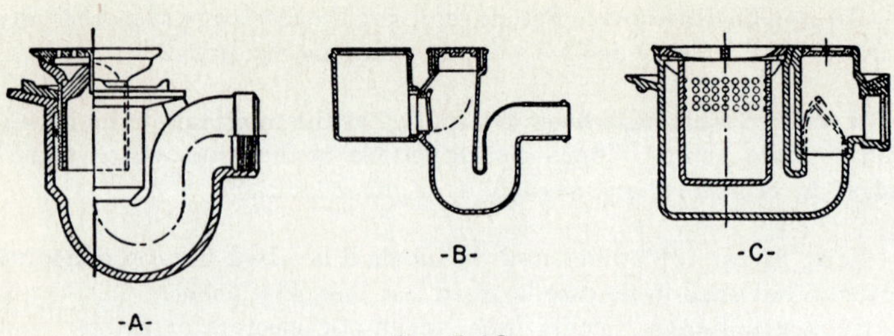

FIG. 5-18

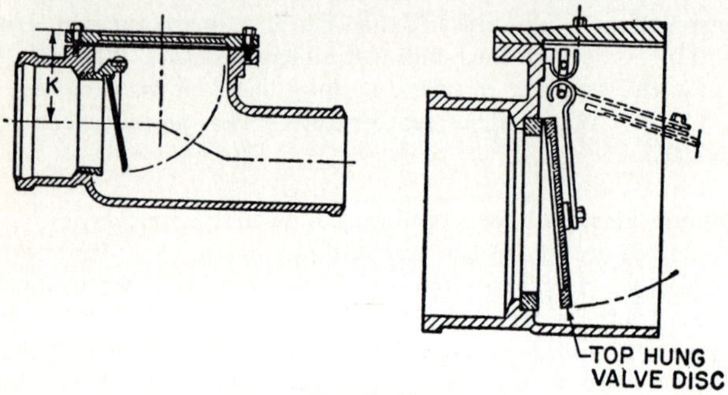

FIG. 5-19

Figure 5-19 illustrates various types of backwater valves that may be installed where the level of the sewer is low and backflow may occur.

Where a floor drain is installed in a seldom-used location, the trap seal might evaporate and permit sewer gas to enter the building. In such case it is well to provide means for refilling the trap seal. Figure 5-20 illustrates one manner whereby a trap seal of a floor drain may be kept full at all times.

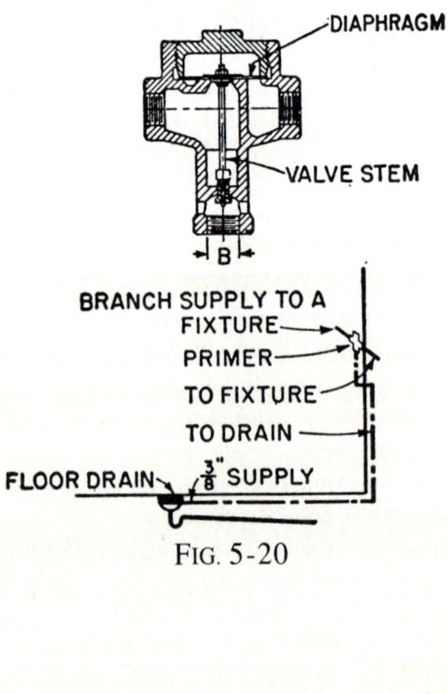

FIG. 5-20

Figure 5-20 shows the means of maintaining a proper seal in basement-floor drain traps and other traps used infrequently. Seldom-used traps tend to lose their seals by evaporation. A primer is installed in a frequently used water-supply line and connected to the trap. Water enters the primer, forces the diaphragm upward, and opens the valve, allowing water to replenish the trap seal. This method prevents foul and unsanitary sewer odors from entering the building.

5.8.4 Size. Floor-drain traps and drains, installed below a basement floor or underground, shall be not less than 2 inches in diameter.

5.8.5 Bell Traps. Bell traps are prohibited.

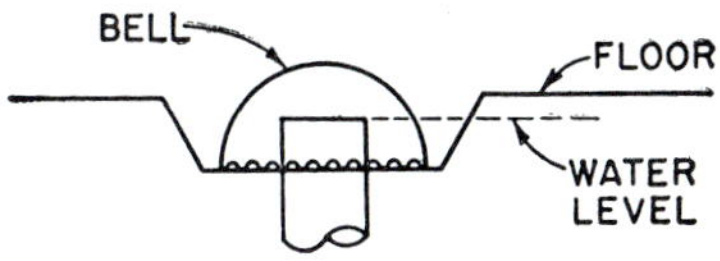

FIG. 5-21

▶ Figure 5-21 shows a bell-trap drain.

5.9 Frost Protection

5.9.1 No soil or waste pipes shall be installed or permitted outside of a building or concealed in outside walls or in any place where they may be subjected to freezing temperatures, unless adequate provision is made to protect them from frost.

▶ **Protection may be by means of frost insulation in cold climates. In climates where the temperature will not be below freezing, there is no mechanical objection to outside installation.**

5.10 Standards for Plumbing Installation Materials.

5.10.1 See Table 5.10.1 for standards applicable to each portion of the drainage system.

Table 5.10.1 Materials for Plumbing Installations

Materials	ASA	ASTM	FS	Other standards remarks
Nonmetallic piping:				
Clay sewer pipe	...	C13-50	SS-P-361a (1942)	Standard strength
		C200-50T	...	Extra strength
Concrete sewer pipe for sizes 4 to 24-in	...	C75-52	...	Reinforced
		C14-52	SS-P-371 (1937)	Nonreinforced
Bituminized-fiber sewer pipe and fittings	...	...	...	Type I CS 116-44[2]
Asbestos-cement sewer pipe	...	...	SS-P-351 (1953)	[1,2]
Ferrous pipe and fittings:				
Cast-iron soil pipe and fittings	A40.1-1935	A74-42	WW-P-401 (1935)[3]	Extra heavy and service weights[2]
Cast-iron water pipe	A21.2-1953	A44-41	WW-P-421 (1931)[4]	AWWA 1908
Cast-iron (threaded) pipe	A40.5-1943	...	WW-P-356 (1936)	
Cast-iron (screwed) fittings	B16.4-1949	...	WW-P-501b (1945)	
Cast-iron drainage fittings	B16.12-1953	...	WW-P-491a (1945)[5]	
Wrought-iron pipe	B36.2-1950	A72-52T	WW-P-441b (1952)[6]	
Steel pipe	...	A120-47	WW-P-406 (1944)	Types I and II
Open-hearth iron pipe	B36.23-1950	A253-51T	WW-P-406 (1944)[7]	Type III only
Malleable-iron fittings	B16.3-1951 (150 lb)	A338-51T*		
Nonferrous pipe and fittings:				
Brass tubing	...	B135-52[8]	WW-T-791-1931	
Brass pipe	H27.1-1953	B43-52	WW-P-351 (1930)[9]	
Brass or bronze flanges and flanged fittings	B16.24-1953 (150 and 300 lb.)			
Cast-brass soldered joint fittings	B16.18-1950	...	...	For copper water tube
Cast-brass solder joint drainage fittings	B16.23-1953			
Bronze screwed fittings, 125 lb	B16.15-1947	...	WW-P-460-(1945)	MSS-SP-10
Copper pipe	H26.1-1949	B42-52	WW-P-377 (1932)	
Seamless copper tubing	...	B75-52	WW-T-797 (1932)[10]	
Copper water tube (KLM)	H23.1-1953	B88-51	WW-T-799a(1943)[11]	SPR 217-49
Wrought-copper and wrought-bronze solder joint fittings	B16.22-1951			
Flared fittings for copper (water) tubes	A40.2-1936			
Lead pipe and traps	...	...	WW-P-325 (1944)	CS 95-41 CS 96-41

Miscellaneous:				
Calking lead	....	....	QQ-L-156(1934)[12] Type I	**CS 94-41**
Sheet lead	....	....	QQ-L-201(1953)[13]	**Grade A**
Sheet brass	....	B36-52 B121-52	QQ-B-611a(1938)	
Sheet copper	....	B152-52	QQ-C-501a(1941)[14]	
Galvanized-iron and steel sheets	G8.8-1937	A163-39	QQ-I-716(1942)[15]	
Galvanized pipe and fittings	....	A93-52T	WW-P-406(1944)	Sec. 18.6
Cement lining	A21.4-1939	A120-47	WW-P-406(1944)	Sec. 18.7
Coal-tar enamel (protective coating)	....	....	....	AWWA 7A.6-1940
Soft solder	....	B32-49	QQ-S-571b(1947)	
Fixture-setting compound	....	....	HH-C-536(1936)	
Air-gap standards	A40.4-1942			
Backflow preventors	A40.6-1943			
Valves:				
Bronze gate	....	....	WW-V-54(1946)[16]	
Cast-iron gate	....	....	WW-V-58(1945)	

* Intended only for use where ASA B16.3 (150 lb) and B16.19 (300 lb) are not adequate.

[1] Asbestos-cement sewer pipe shall meet Federal Specifications SS-P-351 (1940) including Amendment 2, dated Jan. 14, 1942, except for the following substitutions:

Sizes only 4, 5, and 6-inch
Class—nonpressure tests
Lengths: 10 feet—cut of roundness, inside diameter; ±¼ inch
Hydrostatic strength: not applicable
Flexural strength—9-foot span
4-inch pipe—560 lb
5-inch pipe—900 lb
6-inch pipe—1,290 lb
Crushing strength
4-inch pipe—1,740 lb
5 inch pipe—1,680 lb
6-inch pipe—1,420 lb
Tests: one specimen from each 300 lengths of pipe.

[2] See Code chapters for limits of recommended usage.

[3] Amendment 4, dated July 18, 1951, subject: Pipe and Pipe Fittings, Soil, Cast Iron.

[4] Amendment 3, dated Apr. 26, 1940, included.

[5] Amendment 1, dated Feb. 7, 1946, included.

[6] Amendment 2, dated Feb. 8, 1943, included.

[7] Amendment 1, dated June 9, 1945, included.

[8] Mercurous nitrate test required.

[9] Errata 1, dated August, 1933, included.

[10] Amendment 1, June 27, 1946.

[11] Amendment 1, dated Nov. 18, 1946, included.

[12] Amendment 1, dated June 2, 1942, included.

[13] Amendment 3, dated May 27, 1942, included.

[14] Amendment 3, dated November 1948, included.

[15] Amendment 1, dated Apr. 19, 1946, included.

[16] Amendment 1, dated Sept. 30, 1946, included.

Section **6**

Storm Drains

A variety of areas associated with buildings and plants require storm drainage. Such areas include yards, courts, roads, roofs, paved areas, courtyards, etc. Runoff from these areas, which is most often rainwater, is drained into a storm sewer or a combined sewer system.

This section covers the design of storm drains so they are of adequate capacity and properly sloped and installed. The sizes of vertical leaders serving roofs is also discussed. Combined drains and sewers are evaluated, as is continuous flow.

Using the data provided, a designer should be able to plan and size storm drains for any plumbing situation encountered.

6.1 General

6.1.1 Drainage Required. Roofs, paved areas, yards, courts, and courtyards shall be drained into a storm-sewer system or a combined sewer system where such systems are available.

▶ **To comply with this paragraph it is important to compute the total amount of rainfall that might possibly fall over the area to be drained.**

It is essential that the rain water be removed as rapidly as possible. Roof areas should be drained as quickly as the rain falls in order that undue stresses are not placed on the structure.

6.1.2 Prohibited Drainage. Storm water shall not be drained into sewers intended for sanitary sewage only.

▶ **Most modern sewer installations in a city have separate systems: one for the sanitary sewer, which is discharged into the sewage plant, and one for the storm sewer, which is generally disposed of as clear water into a waterway.**

6.1.3 Traps. Leaders and storm drains, when connected to a combined sewer, shall be trapped.

▶ **Traps are required to prevent sewer odors and gases from discharging where they may be offensive.**

If the leader box were located on the roof like the vent terminal so that offensive odors cannot enter the building, there would be no need for trapping. Actually the leaders would help to maintain circulation.

6.1.4 Expansion Joints. Expansion joints or sleeves shall be provided where warranted by temperature variations or physical conditions.

▶ **This requirement is applicable to tall buildings where the movement of the pipe could be sufficient to damage the roof.**

Expansion and improper anchoring of vertical pipes have caused roof drains to be pushed clear above the level of the roof and cause a break on the waterproofing and flashing.

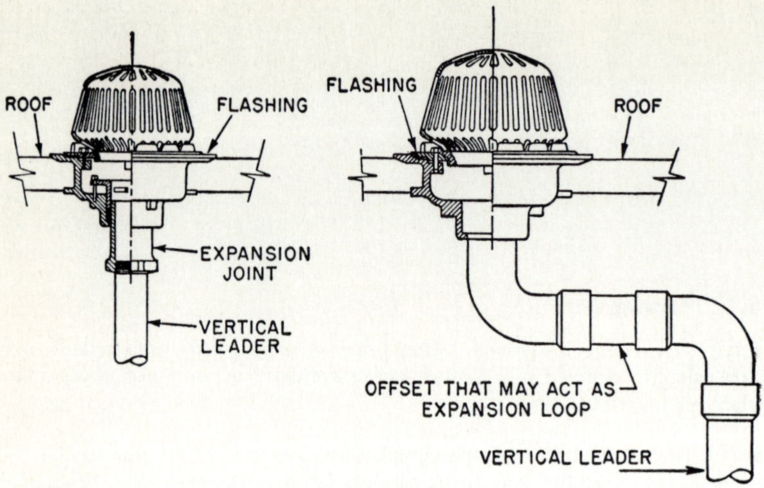

FIG. 6-1

Expansion may be taken care of by a properly designed roof expansion joint or horizontal branch which can take up the movement of the vertical pipe without affecting the roof box. See Fig. 6-1.

6.1.5 Subsoil Drain. Where subsoil drains are placed under the cellar or basement floor or are used to surround the outer walls of a building, they shall be made of open-jointed or horizontally split or perforated clay tile, or perforated bituminized-fiber pipe, or asbestos-cement pipe, not less than 4 inches in diameter. When the building is subject to backwater, the subsoil drain shall be protected by an accessibly located backwater valve. Subsoil drains may discharge into a properly trapped area drain or sump. Such sumps do not require vents.

▶ **Subsoil drains serving outside seepage water or underground soils should be run into a clear-water sump.**

Sumps should be mechanically operated. For a larger building or where it is not possible to watch the sumps, they should be of the automatic type.

In rural areas and for small buildings, subsoil drains may be connected to dry wells or discharged where the discharge will not harm adjacent property.

6.1.6 Building Subdrains. Building subdrains located below the public sewer level shall discharge into a sump or receiving tank, the contents of which shall be automatically lifted and discharged into the drainage system as required for building sumps.

▶ **At times these subdrains might receive waste containing oil or gasoline or other inflammable liquids. In such cases, the discharge must go through an oil interceptor before it is collected into a sump or before discharging into any other disposal system.**

6.2 Materials

6.2.1 Inside Conductors. Conductors placed within a building or run in a vent or pipe shaft shall be of cast iron, galvanized steel, galvanized wrought iron, galvanized ferrous alloys, brass, copper, or lead.

▶ **Any dependable piping materials would be satisfactory. It is in this system of piping that new kinds of piping could be tested before being accepted for the drainage system, especially if the piping is exposed so that it can be watched.**

6.2.2 Outside Leaders. When outside leaders are of sheet metal and connected with a building storm drain or storm sewer, they shall be connected to a cast-iron drain extending above the finish grade, or the sheet-metal leader shall be protected against injury.

6.2.3 Underground Storm Drains. Building storm drains underground, inside the building, shall be of cast-iron soil pipe.

▶ **Outside leaders may be either metallic or nonmetallic. The lower section should be protected from damage by a heavy metal cover or a cast-iron boot.**

Underground storm drains may be the same kind of piping as the building drain.

6.2.4 Building Storm Drains. Building storm drains underground, beneath the building, when not connected with a sanitary or combined sewer, shall be of cast-iron soil pipe or ferrous-alloy piping except that, when found to be adequate by the administrative authority, vitrified-clay pipe, concrete pipe, bituminized-fiber pipe, and asbestos-cement pipe may be used.

▶ **In the preparation of a plumbing code, the local plumbing-code committee decides which kinds of piping are suitable for their area, and only the approved kinds need be spelled out as part of the local code.**

6.2.5 Building Storm Sewers. The building storm sewer shall be of cast-iron soil pipe, vitrified-clay pipe, concrete pipe, bituminized-fiber pipe, or asbestos-cement pipe.

6.3 Traps

6.3.1 Main Trap. Individual storm-water traps shall be installed on the storm-water drain branch serving each conductor, or a single trap shall be installed in the main storm drain just before its connection with the combined building sewer, main drain, or public sewer.

▶ **When the storm leaders are connected into one storm building drain and then connected into the sanitary building drain or storm sewer, a building trap should be installed in order to prevent sewer gases entering the storm leaders.**

If, however, the roof-drain boxes are so located as to preclude any possibility of sewer-gas odors, there is no technical reason to install the building storm trap.

6.3.2 Material. Storm-water traps, when required, shall be of cast iron.

▶ **Traps for the storm drains may be of cast iron or of other composites that have proved safe for the drainage system.**

6.3.3 No traps shall be required for storm-water drains which are connected to a sewer draining storm water exclusively.
6.3.4 Traps for individual conductors shall be the same size as the horizontal drain to which they are connected.
6.3.5 Conductor traps shall be so located that an accessible cleanout may be installed on the building side of the trap.

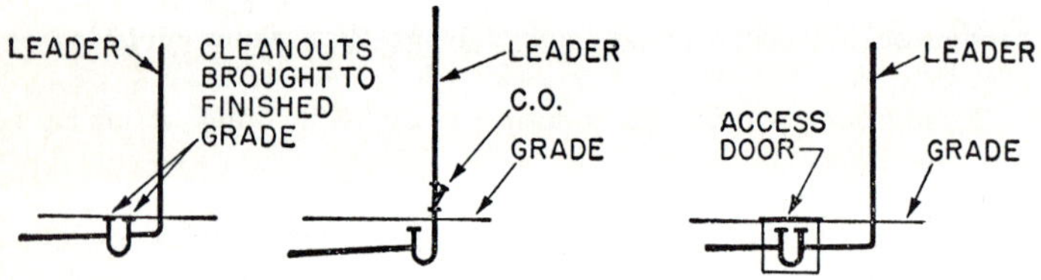

FIG. 6-2

▶ Figure 6-2 illustrates cleanouts for storm leaders.

6.4 Conductors and Connections

6.4.1 Conductor pipes shall not be used as soil, waste, or vent pipes, nor shall soil, waste, or vent pipes be used as conductors.
6.4.2 Rain-water conductors installed along alleyways, driveways, or other locations where they may be exposed to damage shall be protected by metal guards, recessed into the wall or constructed from ferrous-alloy pipe.
6.4.3 Combining Storm with Sanitary Drainage. The sanitary and storm-drainage system of a building shall be entirely separate, except that where a combined sewer is available the building storm drain may be connected in the same horizontal plane through a single wye fitting to the combined drain or sewer at least 10 feet downstream from any branch to the building drain or from any soil stack.

▶ **In certain cities it is permissible to connect both a sanitary and storm sewer into a combined street sewer. As populations increase and cities become more congested, this arrangement is becoming less desirable.**

Figure 6-3 indicates how storm drains may be connected in locations which permit this combined use of the sewer. Some cities require that the storm drains discharge at the curb through a trench or pipe under the sidewalk. Separation of storm and sanitary drains is preferable.

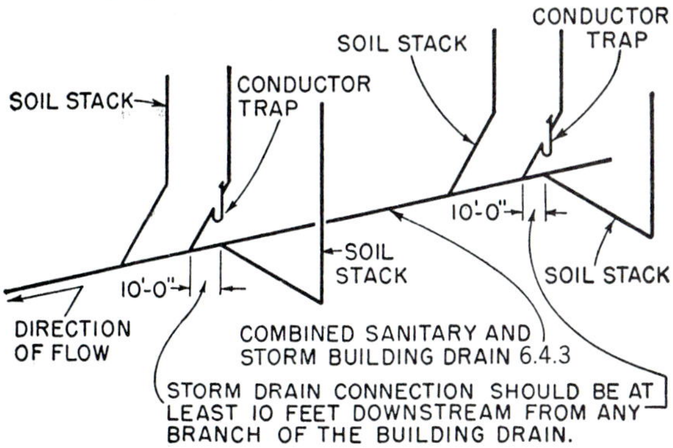

FIG. 6-3

6.4.4 Double Connections of Storm Drains. Where the sanitary and storm drains are connected on both sides of the combined sewer, single wyes shall be used and the requirements of paragraph 6.4.3 relative to the location of connections shall also apply.

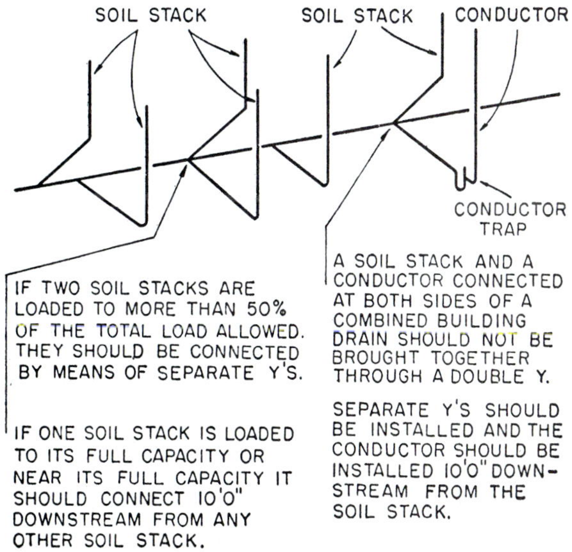

FIG. 6-4

▶ **During a heavy rainstorm the conductor pipe will run almost to its full capacity and may cause pressures to develop on the soil-stack branch unless separated at least 10 feet. This is equally true of a soil stack carrying a relatively heavy load; see Fig. 6-4. If a 4-inch soil stack is loaded to 300 fixture units or more, it represents a relatively heavy load. If the same stack is loaded to 400 fixture units or more, it represents a full load.**

6.4.5 Floor drains connected to a storm drain shall be trapped.

▶ **When floor drains are connected to storm drains, putrefaction of waste matter entering and lodging in the storm drains causes odors, unless floor drains are trapped.**

6.5 Roof Drains

6.5.1 **Material.** Roof drains shall be of cast iron, copper, lead, or other acceptable corrosion-resisting material.

6.5.2 **Strainers.** All roof areas, except those draining to hanging gutters, shall be equipped with roof drains having strainers extending not less than 4 inches above the surface of the roof immediately adjacent to the roof drains. Strainers shall have an available inlet area, above roof level, of not less than 1½ times the area of the conductor or leader to which the drain is connected.

6.5.3 **Flat Decks.** Roof-drain strainers for use on sun decks, parking decks, and similar areas, normally serviced and maintained, may be of the flat-surface type, level with the deck, and shall have an available inlet area not less than two times the area of the conductor or leader to which the drain is connected.

6.5.4 **Roof-drain Flashings.** The connection between roofs and roof drains which pass through the roof and into the interior of the building shall be made watertight by the use of proper flashing material.

▶ **Figure 6-5 illustrates various types of roof drains. Selection of design should be based on the type of roof.**

Mushroom strainers provide a large elevated area for flat roofs where leaves and other debris may accumulate.

Flat strainers are necessary on a roof that is to be utilized as a sun deck or roof garden.

Sloped strainers and corner strainers are necessary when the drain is located in a patched roof or where the drain is placed at the corner of a roof and parapet.

In high structures, roof drains should be provided with expansion joints in order to compensate for structural or temperature variation, unless other means are provided for expansion.

The diameter or size of the drain outlet should be based on the rate of rainfall in the area to be drained.

The open area of the strainer should be 1½ to 2 times the area of the pipe to which it connects.

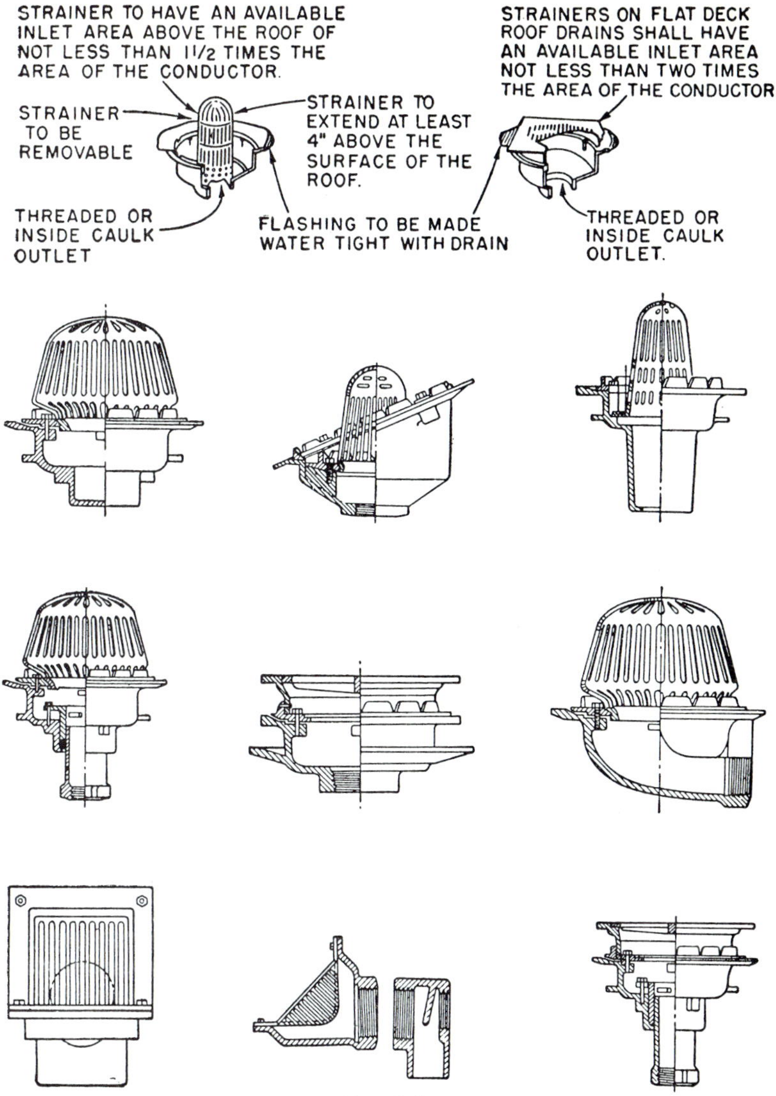

STRAINER TO HAVE AN AVAILABLE INLET AREA ABOVE THE ROOF OF NOT LESS THAN 1 1/2 TIMES THE AREA OF THE CONDUCTOR.
STRAINER TO BE REMOVABLE
STRAINER TO EXTEND AT LEAST 4" ABOVE THE SURFACE OF THE ROOF.
THREADED OR INSIDE CAULK OUTLET
FLASHING TO BE MADE WATER TIGHT WITH DRAIN
STRAINERS ON FLAT DECK ROOF DRAINS SHALL HAVE AN AVAILABLE INLET AREA NOT LESS THAN TWO TIMES THE AREA OF THE CONDUCTOR
THREADED OR INSIDE CAULK OUTLET.

FIG. 6-5

6.6 Size of Leaders and Storm Drains

6.6.1 Vertical leaders shall be sized on the maximum projected roof area, according to Table 6.6.1.

Table 6.6.1 Size of Vertical Leaders

Diam of leader or conductor[1] (in.)	Max projected roof area (sq ft)	Diam of leader or conductor[1] (in.)	Max projected roof area (sq ft)
2	720	5	8,650
2½	1,300	6	13,500
3	2,200	8	29,000
4	4,600		

[1] The equivalent diameter of a square or rectangular leader may be taken as the diameter of that circle which may be inscribed within the cross-sectional area of the leader.

▶ **Table A is an expansion of Table 6.6.1, showing the sizes of vertical leaders for rainfalls of 2 to 8 inches.**

Table A. Expansion of Table 6.6.1 Vertical Leaders

Diam of leader (in.)	Normal rate of rainfall (in.)					
	2	3	4	5	6	8
	Square feet of roof area					
2	1,440	960	720	576	480	360
2½	2,600	1,733	1,300	1,040	865	650
3	4,400	2,933	2,200	1,760	1,470	1,100
4	9,200	6,133	4,600	3,680	3,070	2,300
5	....	....	8,650	6,920	5,765	4,325
6	....	....	....	....	9,000	6,750

6.6.2 Building Storm Drain. The size of the building storm drain or any of its horizontal branches having a slope of ½ inch or less per foot shall be based upon the maximum projected roof area to be handled according to Table 6.6.2.

▶ **Tables 6.6.1 and 6.6.2 are based upon a maximum rate of rainfall of 4 inches per hour. If in any state, city, or other political subdivision, the maximum rate of rainfall is more or less than 4 inches per hour, then the figures for roof area must be adjusted proportionately by multiplying the figure by 4 and dividing the maximum rate of rainfall in inches per hour.**

Table 6.6.2 Size of Horizontal Storm Drains

Diam of drain (in.)	Maximum projected roof area for drains for various slopes (sq ft)		
	1/8 in.	1/4 in.	1/2 in.
3	822	1,160	1,644
4	1,880	2,650	3,760
5	3,340	4,720	6,680
6	5,350	7,550	10,700
8	11,500	16,300	23,000
10	20,700	29,200	41,400
12	33,300	47,000	66,600
15	59,500	84,000	119,000

The U.S. Weather Bureau maintains statistics on the normal rate of rainfall in various localities and will furnish the information on request.

Table B, an expansion of Table 6.6.2, shows the sizes of horizontal storm drains for rainfalls of 2 to 6 inches.

Table B. Expansion of Table 6.6.2 Horizontal Storm Drains

Drain size (in.)	1/8-in. slope, inches rainfall					1/4-in. slope, inches rainfall				
	2	3	4	5	6	2	3	4	5	6
	Square feet of roof area									
3	1,644	1,096	822	657	548	2,320	1,546	1,160	928	773
4	3,760	2,506	1,880	1,504	1,253	5,300	3,533	2,650	2,120	1,766
5	6,680	4,453	3,340	2,672	2,227	9,440	6,293	4,720	3,776	3,146
6	10,700	7,133	5,350	4,280	3,566	15,100	10,066	7,550	6,040	5,033
8	23,000	15,333	11,500	9,200	7,600	32,600	21,733	16,300	13,040	10,866

Figure 6-6 shows approximately equivalent sizes of round and rectangular leaders. In computing the carrying capacity of leaders, there are three factors to consider: (1) dimensions, (2) cross-sectional area, and (3) inside perimeter (area of water contact).

A rectangular leader, because of its four walls and corners, offers greater friction loss, thereby diminishing its carrying capacity. To compensate for this loss, a rectangular leader needs to be about 10 per cent larger than a round leader to carry the same load.

In Fig. 6-6 the sizes for rectangular leaders which are shown as equivalents of the sizes of round leaders include the 10 per cent adjustment. In some cases, the rectangular sizes are more than 10 per cent larger. **Where the computation of the equivalent size has resulted in an unavailable size, the next larger stock size has been given.**

ROUND LEADER			RECTANGULAR LEADER		
DIAMETER INCHES	CROSS SECTIONAL AREA, SQ. IN.	WATER CONTACT AREA, INCHES	DIMENSION INCHES	CROSS SECTIONAL AREA, SQ. IN.	WATER CONTACT AREA, INCHES
2	3.14	6.28	2 X 2 1½ X 2½	4 3.75	8 8
3	7.07	9.42	2 X 4 2½ X 3	8 7.50	12 11
4	12.57	12.57	3 X 4¼ 3½ X 4	12.75 14.0	14.5 14
5	19.06	15.07	4 X 5 4½ X 4½	20.0 20.25	18 18
6	28.27	18.85	5 X 6 5½ X 5½	30.0 30.25	22 22

CIRCUMFERENCE AS STRAIGHT LINE

RECTANGLE AS STRAIGHT LINE

FIG. 6-6

6.6.3 Roof Gutters. The size of semicircular gutters shall be based on the maximum projected roof area, according to Table 6.6.3.

Table 6.6.3 Size of Gutters

Diam of gutter[1] (in.)	Maximum projected roof area for gutters of various slopes (sq ft)			
	1/16 in.	1/8 in.	1/4 in.	1/2 in.
3	170	240	340	480
4	360	510	720	1,020
5	625	880	1,250	1,770
6	960	1,360	1,920	2,770
7	1,380	1,950	2,760	3,900
8	1,990	2,800	3,980	5,600
10	3,600	5,100	7,200	10,000

[1] Gutters other than semicircular may be used provided they have an equivalent cross-sectional area.

▶ Figure 6-7 illustrates comparative sizing for semicircular and rectangular gutters **which aid in computing sizes. The method of selecting sizes is similar to that used for circular and rectangular leaders.**

SEMI-CIRCULAR GUTTERS			RECTANGULAR GUTTERS		
DIAMETER INCHES	CROSS SECTIONAL AREA, SQ. IN.	WATER CONTACT AREA, INCHES	DIMENSION INCHES	CROSS SECTIONAL AREA, SQ. IN.	WATER CONTACT AREA, INCHES
3	3.53	4.70	1 1/2 X 2 1/2	3.75	5.5
4	6.28	6.28	2 1/4 X 3	6.75	7.5
5	9.82	7.85	4 X 2 1/2 3 X 3 1/2	10 10	9 9 1/2
6	14.14	9.43	3 X 5	15	11
8	25.27	12.57	4 1/2 X 6	27	15
10	39.77	15.7	5 X 8 4 X 10	40 40	18 18
		SEMI-CIRCLE AS STRAIGHT LINE	4" 10" 4"		RECTANGLE AS STRAIGHT LINE

FIG. 6-7

6.7 Size of Combined Drains and Sewers

6.7.1 Conversion of roof area to fixture units of storm drainage system that may be connected to a combined sewer. The drainage area may be converted to equivalent fixture-unit loads.

6.7.2 When the total fixture-unit load on the combined drain is less than 256 fixture units, the equivalent drainage area in horizontal projection shall be taken as 1,000 square feet.

6.7.3 When the total fixture-unit load exceeds 256 fixture units, each fixture unit shall be considered the equivalent of 3.9 square feet of drainage area.

6.7.4 If the rainfall to be provided for is more or less than 4 inches per hour, the 1,000-square-foot equivalent in paragraph 6.7.2 and the 3.9 in paragraph 6.7.3 shall be adjusted by multiplying by 4 and dividing by the rainfall in inches per hour to be provided for.

▶ **When a combined storm and sanitary sewer is to be computed, the area of the roof (in square feet) should be converted to fixture units in order to get the total load in fixture units.**

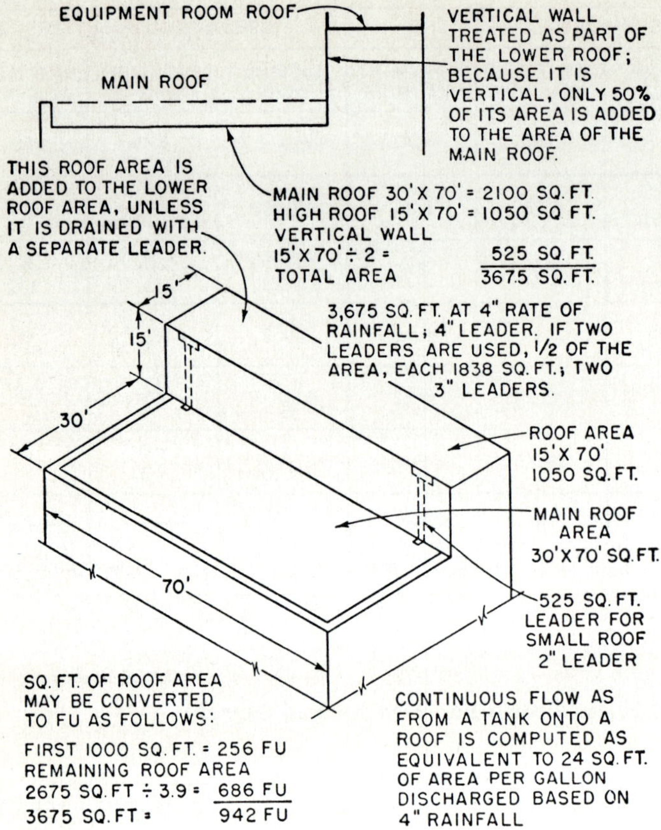

FIG. 6-8

Example: (Fig. 6-8.) Assume that 4,600 square feet of roof area on the storm-drainage system is to be connected into a sanitary drain carrying 700 fixture units.

	Fixture Units
The first 1,000 sq ft	256
Remaining 3,600 sq ft divided by 3.9	923
Subtotal 4,600 sq ft	1,179
Add sanitary drainage system	700
Grand total	1,879

According to Table 5.5.2, 1,879 fixture units require a 10-inch diameter drain.

The conversion of roof area to fixture units is based on a 4-inch rate of rainfall. If a different rate of rainfall is used, the computation changes proportionately.

6.8 Values for Continuous Flow

6.8.1 Where there is a continuous or semicontinuous discharge into the building storm drain or building storm sewer, as from a pump, ejector, air-conditioning plant, or similar device, each gallon per minute of such discharge shall be computed as being equivalent to 24 square feet of roof area, based upon a 4-inch rainfall.

▶ Allow for the extension of walls above the building and the amount of rain that should be considered for runoff of the roof area itself. Figure 6-8 illustrates this condition for a flat roof. The roof area would then be 70 by 30 feet, or 2,100 square feet. The wall area is 15 by 70 feet, or 1,050 square feet. However, the wall area would not represent 100 per cent of the required drainage area. One-half of the wall area is sufficient in calculating the proper drainage. Add 2,100 square feet of roof area plus half of the wall area (525 square feet), or a total of 2,625 square feet. According to Table 6.6.1, 2,625 square feet require a 4-inch-diameter conductor based on a normal rainfall of 4 inches per hour.

In the National Plumbing Code, after Tables 6.6.1 and 6.6.2, is given a formula for adjusting the roof area in square feet to rates of rainfall other than 4 inches.

The following is a formula for proportioning the sizes of leaders or drains to rates of rainfall other than 4 inches:

Example a for 5-inch Rainfall: Table 6.6.1 prescribes a 3-inch leader for a 2,200-square-foot roof area in a locality of 4-inch rainfall. For 5-inch rainfall, this leader would be too small, as it could handle only $\frac{4}{5}$ of the rainfall. To figure the correct size, invert $\frac{4}{5}$ to $\frac{5}{4}$ and multiply by 3: $5 \times 3/4 = 3.75$-inch leader. As this is not a standard size, use the next standard size, 4-inch.

Example b for 3-inch Rainfall: Table 6.6.2 prescribes a 6-inch storm drain for a 7,500-square-foot roof area at $\frac{1}{4}$-inch slope in a locality of 4-inch rainfall. For 3-inch rainfall, this drain would be too large, as it could handle $\frac{4}{3}$ of the load. To figure the correct size, invert $\frac{4}{3}$ to $\frac{3}{4}$ and multiply by 6: $3 \times 6/4 = 4.5$-inch drain. Use the next standard size, 5 inch.

Section **7**

Sizing Drainage Systems

Drainage systems must be capable of handling the anticipated flow without overloading. At the same time, the building or facility owner does not want to invest in an overly large system that costs more than necessary. It is the designer's task to see that the system meets both needs, capacity and cost, while being completely reliable.

This section provides the designer with the techniques needed to plan and design drainage systems that are adequate in size and of acceptable cost. Using the data provided, a designer can produce satisfactory designs for any type of plumbing installation. Commonly accepted code requirements are included.

SIZING FOR ADEQUACY OF PERFORMANCE

This section deals with the application of principles of hydraulics in sizing drainage piping so as to achieve the performance required of plumbing fixtures, fixture drains, sanitary drainage stacks, branches and building drains, storm drainage systems, and combined storm and sanitary building drains.

The specific criteria used in each sizing application are presented in detailed discussion for each section of drainage piping systems. The criteria define the limits for adequacy of performance.

FIXTURE DRAINS

In establishing proper sizes for fixture drains, consideration should be given to the diameter of the orifice in the waste-outlet fitting and tailpiece, the size of the trap, and the desired velocity of flow for proper fixture drainage and scouring action in the fixture drain. Fixtures generally have discharge orifices sized to provide suitable rates of drainage. In no case should the fixture drain size be reduced to less than that of the discharge orifice, for to do so would cause a reduction in the rate of fixture discharge. Instead, the size of the fixture trap and the fixture drain should be slightly larger than that of the discharge orifice so that they will not affect the discharge rate adversely after a moderate degree of fouling in normal service. Similarly, the size of the fixture drain should not be smaller than that of the fixture trap, so as to avoid constriction of flow and resultant increased fouling in the fixture trap.

Where fixtures discharge greasy wastes, low velocity of flow in fixture drains increases the incidence of stoppage development therein. For such fixtures, it is recommended that the size of the fixture trap and drain should be the minimum commercial size available and larger than the size of the discharge orifice. In this way, maximum scouring effect may be achieved to minimize fouling of the fixture drain.

Hence, minimum fixture drain sizes may be assumed to be the same as the minimum sizes permitted for the fixture traps of the various kinds of plumbing fixtures.

SANITARY DRAINAGE FIXTURE UNITS

A method of assigning fixture unit values to plumbing fixtures was originally proposed in 1923 by Roy B. Hunter of the National Bureau of Standards, U.S. Department of Commerce. It was presented in the publication, "Recommended Minimum Requirements for Plumbing," Department of Commerce Building and Housing Series, BH13, 1924. The suggested values were designed for application in conjunction with the probability of simultaneous use of fixtures so as to establish maximum permissible drainage loads, in terms of fixture units rather than in numbers of specific types of fixtures or gallons per minute of drainage flow, for each of the various parts of sanitary drainage systems. Since the original proposal, various changes were recommended and made by Hunter, in a 1940 National Bureau of Standards publication, BMS 65, and more recently by others participating in development of current model plumbing regulations.

Fixture unit values presently recommended for assignment to various kinds of plumbing fixtures which discharge into sanitary drainage systems are stated in Table 7-1. They are provided as a means for computing sizes of soil, waste, and vent piping based upon the loading effects produced by the discharge of many different kinds of plumbing fixtures commonly installed in buildings.

In general, the sanitary drainage fixture unit value assigned to a particular fixture is based upon the average volume discharged and the average rate of discharge for the fixture. This value is determined from the fixture's total discharge flow, in gallons per minute, divided by 7.5; or in other words, its total discharge flow in cubic feet per minute. The integer closest to the flow rate, in cubic feet per minute, is taken as the fixture unit value. Thus, the sanitary drainage fixture unit is a factor so chosen that the drainage load producing effects of different plumbing fixtures are expressed approximately as multiples of that factor.

Table 7-1
SANITARY DRAINAGE FIXTURE UNIT VALUES

Fixture or group	Trap size in	Trap size mm	Fixture units
Residential:			
Automatic clothes washer, domestic	2	50.0	3
Bathroom group consisting of a water closet, lavatory, and bathtub or shower stall:			
Flushometer valve closet			8
Tank-type closet			6
Bathtub (with or without overhead shower)	1½	37.5	2
Bidet	1¼	31.3	1
Dishwasher, domestic	1½	37.5	2
Floor drain	2	50.0	3
Floor drain	3	75.0	5
Floor drain	4	100.0	6
Food waste grinder, domestic	1½	37.5	2
Kitchen sink, domestic	1½	37.5	2
Kitchen sink, domestic, with dishwasher	1½	37.5	2
Kitchen sink, domestic, with food waste grinder	1½	37.5	2
Kitchen sink, domestic, with dishwasher and food waste grinder	2	50.0	2
Kitchen sink and wash (laundry) tray with single 1½-in (37.5-mm) trap	1½	37.5	2
Kitchen sink and wash (laundry) tray with separate 1½-in (37.5-mm) traps	1½	37.5	3
Kitchen sink and wash (laundry) tray with food waste grinder unit	2	50.0	4
Lavatory, common	1¼	31.3	1
Laundry tray (1 or 2 compartments)	1½	37.5	2
Shower stall, single head	2	50.0	2
Sink, bar, private	1½	37.5	1
Water closet, tank-type, trap arm only	3	75.0	4
Public toilet rooms:			
Urinal, pedestal, trap arm only	3	75.0	6
Urinal, pedestal, siphon jet blowout	3	75.0	6
Urinal, stall, washout	2	50.0	4
Urinal, wall [2-in (50-mm) min. waste]	1½	37.5	4
Water closet, Flushometer valve, trap arm only	3	75.0	6

Table 7-1 (*Continued*)

Fixture or group	Trap size in	Trap size mm	Fixture units
Industrial:			
Interceptors for grease, oil, solids, etc.	2	50.0	3
Interceptors for sand, auto wash, etc.	3	75.0	6
Lavatory, multiple-type (wash fountain or wash sink)	1½	37.5	2
Showers, gang (one unit per head)	2	50.0	2
Commercial:			
Dishwasher, commercial	2	50.0	2
Food waste grinder, commercial	2	50.0	3
Receptors (floor sinks) indirect waste receptors for refrigerators, coffee urn, water stations, etc.	1½	37.5	1
Receptors, indirect waste receptors for commercial sinks, dishwashers, airwashers, etc.	2	50.0	3
Sink, bar, commercial [2-in (50-mm) min. waste]	1½	37.5	2
Sink, commercial, with food waste grinder	2	50.0	3
Sink, commercial (pot, scullery, or similar type)	2	50.0	4
Sink (flushing-rim type, flush valve supplied)	3	75.0	6
Sink (service type with trap standard)	3	75.0	3
Sink (service with P-trap)	2	50.0	2
Washing machines, commercial	2	50.0	3
Medical:			
Dental unit or cuspidor	1¼	31.3	1
Dental lavatory	1¼	31.3	1
Lavatory (surgeon's, barber shop, beauty parlor)	1½	37.5	2
Sink (surgeon's)	1½	37.5	3
Miscellaneous:			
Drinking fountain	1¼	31.3	½
Mobile home park traps (one for each trailer)	3	75.0	6
Trap size 1¼ in (31.3 mm) or less	1¼	31.3	1
Trap size 1½ in (37.5 mm)	1½	37.5	2
Trap size 2 in (50 mm)	2	50.0	3
Trap size 2½ in (62.5 mm)	2½	62.5	4
Trap size 3 in (75 mm)	3	75.0	5
Trap size 4 in (100 mm)	4	100.0	6

DESIGN LOAD FOR SANITARY DRAINAGE SYSTEM

Except in the case of fixtures in which water is retained for use and discharged afterward, the rate at which water leaves fixtures and enters the sanitary drainage system is the same as the rate at which water is supplied to the fixtures. For relatively large systems where fixtures are supplied directly from public water mains, the rate at which water enters the water supply distributing system and the rate at which sewage leaves the building drainage system are the same except for a small amount of water consumed by occupants. Hence, there is relative equivalency between the water supply demand load and the sanitary drainage discharge load for which the respective systems are designed.

This ultimate equivalency is the basic premise upon which values have been assigned to express the load producing effects of fixtures, both in water supply fixture units and in sanitary drainage fixture units. Although the values assigned to fixtures for water supply purposes differ from those for sanitary drainage, the same design load in gallons per minute corresponding to a given number of fixture units is applicable to the main lines of both systems, with one proviso. The design load for sanitary drainage systems should be determined from the number of sanitary drainage fixture units and the corresponding water supply demand load, in gallons per minute as indicated in the charts (Fig. 3-3) and in Table 3-2, applicable to systems equipped predominantly with flush valves (Flushometers). The reason for this is simply that the discharge rate of water closets equipped with flush tanks is sufficiently comparable with that of flush-valve equipped water closets as to warrant use of the demand rates applicable to systems equipped with flush valves (Flushometers).

DRAINAGE STACKS AND BRANCHES

Based on the computed drainage stack flow capacity, for stacks flowing $7/24$ full at terminal velocity, the corresponding number of fixture units may be determined from design load charts or tables, so as to establish the total load which may be placed on a tall drainage stack. For example, the computed flow capacity of a 4-in (100-mm) stack flowing at $7/24$ full is 143 gpm (9.02 L/s). From design load charts or tables, it may be found that this rate of flow is equivalent to 500 fixture units.

This is the total load which may be received from all branches on a 4-in (100-mm)-tall stack. However, to avoid excessive interference between flow entering the stack and that coming down the stack, it is necessary to limit the amount of flow which may be allowed to enter the stack at each of the branches. Thus, in a building of just a few stories in height, the amount of flow entering the stack through a branch may be greater than would be permissible in a building of many stories.

Table 7-2 for sizing drainage stacks provides different permissible loadings for stacks of three stories or less in height and for stacks more than three stories in height. Included in the table are the maximum loads permitted on any horizontal fixture branch of a short stack and at any one story of a stack more than three stories in height.

SANITARY BUILDING DRAINS

Sanitary building drains should be designed to flow only partially full so as to provide an air space in the upper portion of the pipe through which air may flow and relieve pressure therein to available branches. A general recommendation in this case is to design such drains to flow at from one-half to three-quarters of the quantity rate of discharge which may be calculated for uniform flow conditions in sloping horizontal sanitary drains flowing full.

In relatively low buildings of extensive area, the condition of flow in the sanitary building drain may approach uniformity within the building, and for such buildings it is recommended that the drain be designed to convey liquid flow at one-half of its full-flow capacity under such conditions. This permits the drain to flow at one-half depth for maximum liquid load.

In buildings of the multistory type, the condition of flow in the sanitary building drain usually is different because of the high velocity at which discharges are received from drainage stacks. Considerable turbulence and surging flow occur in such cases. Under such conditions, flow is conveyed by the building drain at velocities higher than those that prevail for uniform flow conditions. For such buildings, it is recommended that the drain be designed to convey liquid flow at three-fourths of the quantity rate of flow corresponding to full-flow capacity under uniform flow conditions. The higher capacity may be permitted because of the higher velocities of surging flow, and correspondingly higher capacities for drains flowing at one-half depth.

Table 7-2 may be used for sizing sanitary building drains. The value shown therein for an 8-in (200-mm) drain laid at a ¼ in/ft (20.8 mm/m) slope corresponds to a design load, in gallons per minute, equal to the half-full flow rate for such a drain flowing at uniform flow conditions. The values shown for smaller sizes reflect an allowance made for the effects of surging flow. Similar allowances have been made for each given size of drain reflecting the effect of surging flow on the relative capacities of drains at different degrees of slope.

Table 7-2
MAXIMUM PERMISSIBLE LOADS FOR SANITARY DRAINAGE PIPING
In Terms of Fixture Units

Pipe diameter		Any horizontal fixture branch	One stack of 3 stories or less in height	Stacks more than 3 stories in height		Building drain, and building drain branches from stacks — Slope, in/ft (mm/m)			
in	mm			Total for stack	Total at one story	1/16 (5.2)	1/8 (10.4)	1/4 (20.8)	1/2 (41.6)
1½*	37.5	3	4	8	2	np	np	np	np
2*	50	6	10	24	6	np	np	21	26
2½*	62.5	12	20	42	9	np	np	24	31
3	75	20†	48‡	72‡	20†	np	np†	42†	50†
4	100	160	240	500	90	np	180	216	250
5	125	360	540	1100	200	np	390	480	575
6	150		960	1900	350	np	700	840	1000
8	200			3600	600	1400	1600	1920	2300
10	250			5600	1000	2500	2900	3500	4200
12	300					3900	4600	5600	6700

* No water closets permitted.

† Not over two water closets permitted.

‡ Not over six water closets permitted.

STORM DRAINAGE SYSTEM

Storm drains in buildings may be sufficiently long so that flow has ample time to adjust itself and reach a state of equilibrium or uniform flow condition. High velocity and surges in flow entering a long horizontal drain of constant shape and slope are dissipated gradually owing to pipe friction. Thereafter, flow proceeds at a rate corresponding to uniform flow conditions for the drain.

Sizing of horizontal storm drains is based upon the quantity rate of uniform flow for drains flowing full, under no pressure, after many years of service. The velocity and capacity corresponding to full flow in galvanized-iron and cast-iron storm drains, laid at a slope of ¼ in/ft (20.8 mm/m) are shown in Table 7-4. For drains laid at other slopes, appropriate multipliers are also given in the accompanying text. These multipliers may be applied to the values tabulated for drains laid at a ¼ in/ft (20.8 mm/m) slope.

The maximum rate at which storm water should be conveyed from drainage areas may be assumed to be the same as the rate at which storm water may collect thereon. This is dependent upon the maximum rainfall rate recorded in the region of the building installation and upon the amount of storm water collection area. For many regions in the United States, the maximum rainfall rate recorded over a 10-min period is 4 in (100 mm)/h. At this rate, the amount of drainage area upon which 1 gal (3.79 L) of storm water will collect in 1 min is 24 ft^2 (2.23 m^2).

By applying a value of 24 ft^2 (2.23 m^2) as the collection area equivalent to 1 gpm (3.79 L/s) of maximum storm drainage flow, tables showing uniform flow capacities in gallons per minute for sloping drains flowing full can be converted to terms of maximum permissible storm drainage areas which may be served by various sizes of drains in such regions.

Such a table for use in sizing storm drainage piping is given in Table 7-3. In the table, the values specified as maximum permissible drainage areas for horizontal storm drainage piping were derived in the manner discussed above for gravity flow in sloping drains flowing full under uniform flow conditions. For leaders, the values correspond to full-flow capacities of horizontal drains laid at a slope of ¾ in/ft (62.4 mm/m). For semicircular gutters, the values correspond to flow rates which are less than one-half of the full-flow capacities of circular horizontal drains of the same diameters. The 3-in (75-mm) gutter is assigned 29.3 percent, and the 8-in (200-mm) gutter is assigned 24.4 percent, of full-flow capacity for circular horizontal drains for such sizes. These percentages of full-flow capacity correspond, respectively, to flow depths of 37.5 and 35 percent of the diameter of semicircular gutters. Hence, flow at capacity should produce no spillage over the edge of the gutter at peak rainfall rate.

COMBINED STORM AND SANITARY BUILDING DRAINS

Where sanitary building drains are connected to storm building drains so as to form a combined storm and sanitary drain, an equivalency basis is necessary to convert the fixture unit load of the sanitary system to that of the storm drainage system in square feet of drained area. As the flow in the storm drainage system during maximum rainfall periods generally is far in excess of the flow in the sanitary drainage system at peak load, the sizing of a combined drain should be based on the storm drain design, and sanitary drainage fixture units of load should be converted to terms of equivalent square feet of drained area.

A method of converting such load units has been recommended. As a minimum, 1000 ft² (92.9 m²) of drained area should be deemed equivalent to all sanitary drainage fixture unit loads that are less than 256 units. It may be seen from the water supply system demand load charts and Table 3-2 that each additional fixture unit above 256 corresponds to an additional load of 0.2 gpm (0.0126 L/s). At a maximum rainfall rate of 4 in (100 mm)/h, rain is collected at a rate of 1 gpm (0.063 L/s) on 24 ft² (2.33 m²) of drained area and at a rate of 0.2 gpm (0.0126 L/s) on 3.9 ft² (0.36 m²) of drained area.

Consequently, each fixture unit of sanitary drainage load in excess of 256 such units should be deemed equivalent to 3.9 ft² (0.36 m²) of drained area. The total drained area equivalent to the number of sanitary drainage fixture units should be added to the total storm drainage area to determine the load on the combined system, and the appropriate size of combined drain then may be selected from the tables applicable for sizing horizontal storm drains.

STANDARD CODE REGULATIONS

Sanitary Drainage Fixture Units Fixture unit values given in Table 7-1 shall be employed in computing the total load carried by a soil or waste pipe and shall be used with tables for sizing soil, waste, drain and vent pipes, except as otherwise prescribed herein. The total discharge flow in gallons per minute for any single fixture, divided by 7.5, provides the fixture unit value for that particular fixture.

Sizing the Sanitary Drainage System The maximum permissible fixture unit load that may be connected to any horizontal fixture branch, soil stack, waste stack, building drain, and building drain branches from stacks, shall be determined from Table 7-2, except as otherwise provided herein.

Sanitary drainage systems shall have at least one main stack extending undiminished in size and as directly as possible from the building drain to the open air above the roof. Such stack shall be at least 3 in (75 mm) in size, but not larger than the size of the building drain.

Soil and waste stacks shall be at least as large as the largest branch connection thereto, except that it shall be permissible to connect the 4-in (100-mm) fixture drain of a water closet to a 3-in (75-mm) soil stack by means of a 3″ × 3″ × 4″ TY or sanitary tee.

The size of the sanitary building drain at any point shall be at least as large as the largest branch or stack located upstream from such point.

Underground waste piping shall be at least 2 in (50 mm) in size.

Where a soil or waste stack is offset at an angle of more than 45° from the vertical, and the fixture unit load conveyed by the offset exceeds the permissible load for building drains of the same size as the stack, the offset shall be sized as a building drain for that load, and the size of the stack section below the offset shall be at least as large as the offset.

Where drainage or vent piping outlets are provided for future fixture installations, they shall be considered in determining the sizes of drainage and vent piping.

Sizing the Storm Water Drainage System The size of a building storm drain or any of its horizontal branches having a slope of ½ in or less per foot shall be based upon the total drainage area served by such drains in accordance with Table 7-3.

Table 7-3
MAXIMUM PERMISSIBLE LOADS FOR STORM DRAINAGE PIPING AND GUTTERS
In Terms of Square Feet of Projected Drainage Area

Pipe or gutter diameter*			Horizontal storm drainage piping			Horizontal storm drainage gutters			
			Slope, in/ft (mm/m)			Slope, in/ft (mm/m)			
in	mm	Leaders	⅛ (10.4)	¼ (20.8)	½ (41.6)	¹⁄₁₆ (5.2)	⅛ (10.4)	¼ (20.8)	½ (41.6)
2	50	720	np	np	np	np	np	np	np
2½	62.5	1,300	np	np	np	np	np	np	np
3	75	2,200	822	1,160	1,644	170	240	340	480
4	100	4,600	1,880	2,650	3,760	360	510	720	1,020
5	125	8,650	3,340	4,720	6,680	625	880	1,250	1,760
6	150	13,500	5,350	7,550	10,700	960	1,360	1,920	2,720
8	200	29,000	11,500	16,300	23,000	1,990	2,800	3,980	5,600
10	250		20,700	29,200	41,400	3,600	5,100	7,200	10,200
12	300		33,300	47,000	66,600				
15	375		59,500	84,000	119,000				

Note: This table is based upon a rainfall rate of 4 in/h. Where greater rates frequently occur, the areas given in the table shall be adjusted by multiplying them by 4 and dividing by such greater rate, in inches per hour.

* The equivalent diameter of a square or rectangular leader may be taken as the diameter of that circle which may be inscribed within the cross-sectional area of the leader. Gutters other than semicircular may be used provided they have an equivalent cross-sectional area.

The size of subsoil drains shall be not less than 4 in (100 mm) where installed under a cellar or basement floor or where installed surrounding the outer walls of a building.

Leaders shall be sized on the basis of the horizontally projected roof areas served, except that an allowance shall be made for any additional storm water drainage load received by such roof areas from adjacent walls exposed to the elements. Leaders shall be sized in accordance with Table 7-3.

The size of a semicircular gutter shall be based upon the horizontally projected roof area served by such gutter in accordance with Table 7-3.

Sizing Combined Storm and Sanitary Building Drains The size of any combined drain shall be based on a drainage area equivalent to the total sanitary and storm drainage loads served by such drain, and shall be determined as for horizontal storm drains, using Table 7-3. The sanitary drainage load shall be converted to equivalent storm drainage area, and then added to the storm drainage area. Sanitary drainage loads shall be converted as follows:

> Where the total fixture unit load on the combined drain is less than 256 fixture units, the sanitary drainage load shall be taken as equivalent to 1000 ft² (92.9 m²) of storm drainage area.
>
> Where the total fixture unit load on the combined drain is more than 256 fixture units, the sanitary drainage load shall be computed by considering each fixture unit as equivalent to 3.9 ft² (0.36 m²) of storm drainage area.
>
> Where there is a continuous or intermittent flow into the drain or sewer, as from a pump, ejector, air-conditioning equipment or similar equipment, each gallon per minute (0.063 L/s) of flow at rated capacity shall be computed as being equivalent to 24 ft² (2.23 m²) of storm drainage area.

Table 7-4
UNIFORM FLOW VELOCITY AND CAPACITY OF GALVANIZED-IRON AND CAST-IRON DRAINS OF STORM DRAINAGE SYSTEMS AT FULL FLOW
¼ in/ft (20.8 mm/m) Slope

Pipe diameter		Velocity		Capacity	
in	**mm**	**fps**	**m/s**	**gpm**	**L/s**
2	50	1.72	0.52	17.4	1.10
2½	62.5	1.99	0.61	31.5	1.99
3	75	2.25	0.69	51.3	3.24
4	100	2.74	0.84	111	7.00
5	125	3.16	0.96	201	12.7
6	150	3.58	1.09	327	20.6
8	200	4.35	1.33	705	44.5

Section **8**

Sizing Vent Piping

To prevent a vacuum from being developed in any plumbing system in an unwanted manner, vent piping must be used. And this vent piping must be of sufficient size to allow adequate quantities of air or gas to flow.

This section presents the best methods to size vent piping for any type of plumbing system. It covers vent stacks, vent extensions, individual vents, vent headers, branch vents, circuit and loop vents, suds pressure release vents, and vents for building sewage sumps and receiving tanks.

Using the data presented here, a designer can size vents so they function properly in the plumbing system. Standard code regulations are also considered and discussed.

SIZING FOR ADEQUACY OF PERFORMANCE IN EXTENDED SERVICE

This chapter deals with the application of principles of fluid mechanics and pneumatics in sizing vent piping so as to achieve adequacy of performance in extended service.

The specific criteria used in each sizing application are presented in detailed discussion for each section of the attendant vent piping system. The criteria define the limits for adequacy of performance.

VENT STACKS

For any given quantity rate of airflow required to be conveyed by a vent stack, the maximum permissible total length of vent piping of a given diameter may be readily computed by means of a formula [Eq. (8-3)]. The total length determined in this manner should be understood to be the sum of the total developed length of the vent stack, measured from its base connection to its terminal in the atmosphere above the roof of the building, plus the total equivalent pipe length allowable for the pipe fittings in that run of vent piping. In general, the equivalent length of pipe fittings in multistory buildings has been found to work out to be approximately 50 percent of the developed length of vent stacks. Hence, it may be assumed that the maximum permissible developed length is two-thirds of the computed total vent stack length.

It may be noted also that the maximum airflow capacity of a given diameter of vent stack of zero length is that which may be computed as the quantity rate of discharge into the atmosphere from outlets of

Table 8-1
SIZE OF VENT STACKS AND BRANCH VENTS

Size of soil or waste stack, in	Size of soil or waste stack, mm	Fixture units connected	Diameter of vent required, in (mm): 1¼ (31.3)	1½ (37.5)	2 (50)	2½ (62.5)	3 (75)	4 (100)	5 (125)	6 (150)	8 (200)
			Maximum developed length of vent, ft								
1¼	31.3	2	30								
1½	37.5	8	np	150							
2	50	24	np	50	150						
2½	62.5	42	np	np	100	300					
3	75	72	np	np	np	80	400				
4	100	500	np	np	np	np	180	700			
5	125	1100	np	np	np	np	np	200	700		
6	150	1900	np	np	np	np	np	np	200	700	
8	200	3600	np	np	np	np	np	np	np	250	800
10	250	5600	np	np	np	np	np	np	np	np	250

a vent pipe wherein the pressure is 1 in (25 mm) of water column. This may be calculated by means of a formula [Eq. (8-1)].

To relieve suds pressure at the base of drainage stacks, it may be appropriate to connect the base of the vent stack to a point in the horizontal drain directly adjacent to the base fitting of the drainage stack and to size the lowest two-story section of the vent stack in accordance with the size computed by means of a formula [Eq. (8-2)] for suds discharge rate from outlets to the atmosphere. The computed size for suds relief purposes may be found to be at least three-fourths of the drainage stack size.

Table 8-1 may be used for sizing vent stacks in accordance with drainage stack capacity loads. Permissible lengths of vent piping shown therein are sufficiently less than those that may be computed by formulas (in which additional allowance need be made for the equivalent length of pipe fittings) that the stated lengths may be applied directly as permissible developed length of pipe. It should be understood that this table applies exclusively to airflow loads of vent stacks attendant upon drainage stacks and is not intended for application to suds flow conditions.

VENT EXTENSIONS AND TERMINALS OF STACKS

The size of vent extensions and terminals of soil, waste, and vent stacks is directly related to their capacities for permitting air to enter the tops of stacks without causing excessive air pressure reduction in the upper section of the sanitary drainage system. In no case should the size of the vent extension and terminal of a drainage stack be reduced to less than that required for its attendant vent stack, for air must be permitted to enter the top of the drainage stack at the same quantity rate of flow as it is dragged down the stack by water at capacity load.

To achieve maximum effective circulation of air up through the sanitary drainage system during periods of relatively small load, it is recommended that the vent extension and terminal of the drainage stack which connects to the most upstream end of the sanitary building drain be of the same size as the drainage stack. If vent extensions and terminals are not to restrict gravity circulation of air in the system, the sum of the cross-sectional areas of all terminals should be equal to the cross-sectional area of a fresh air inlet through which air is supplied into the sanitary building drain of the system or, where there is no house trap and fresh air inlet on the sanitary building drain, equal to the cross-sectional area of the sanitary building drain at its point of entry into the building.

Where vent extensions and terminals are subject to frost closure conditions during frigid weather, provision should be made in sizing to ensure against the occurrence of excessive reduction of available flow area due to formation of ice within the vent extension or at the terminal. A recommendation in this regard is to provide vent extensions and terminals of at least 3-in (75-mm) size for all smaller stacks and to make such necessary changes in size by installation of a long increaser below the roof line within the building.

Similar recommendations for the sizing of vent extensions and terminals may be found in the section "Standard Code Regulations" in this chapter.

VENT HEADERS, THEIR VENT EXTENSIONS AND TERMINALS

Vent headers should be of adequate size to convey air at the quantity rates of flow required at the tops of each of the drainage stacks served thereby without causing excessive air pressure reduction in the upper section of the sanitary drainage system. When the system is operating at capacity, the quantity rate of air required to be supplied through the vent header to drainage stacks is directly related to the total drainage load of all stacks served by the vent header, rather than being just the sum of the individual airflow requirements for each of the stacks at their respective capacities. Thus, the various sections of a vent header should be sized as for a vent stack serving a drainage stack of capacity sufficient for the total drainage load, the length of the vent stack being the same as that of the longest vent stack connected to the vent header.

Vent extensions and vent terminals through which air is supplied to vent headers from the atmosphere above the building roof should be sized in the same manner as for any other section of a vent header. In this case, the vent extension and terminal convey the greatest quantity of airflow of all sections of the vent header and, hence, should be the largest in size.

A similar basis for sizing vent headers is given in the section "Standard Code Regulations" in this chapter. It provides for sizing sections of a vent header in accordance with the sum of the fixture unit loads of the stacks vented through a section and the developed length of the vent stack having the longest developed length to the open air.

INDIVIDUAL VENTS

Proper sizes for individual vents connected to fixture drains should be determined in accordance with the principal function performed by such vents in any given installation and the corresponding quantity rate of airflow required to be conveyed. This may be established by analysis of the arrangement of drainage and vent piping for any particular system.

Where individual vents are connected to fixture drains which discharge into drainage stacks, or into branch drains of stacks, above the level of the base connection of an attendant vent stack of adequate size, the principal function of the individual vent can be assumed to be that of providing sufficient airflow to the fixture drain so as to disrupt siphonic action and thereby avoid excessive trap seal loss during discharge of the fixture. The quantity rate at which air is required may be determined from Table 8-3, based upon the size of the fixture drain. A corresponding size may be computed by means of Eq. (8-3) and by reference to Table 8-2, based upon the length of vent pipe measured from the fixture drain to the branch fitting in the vent stack or to its individual vent terminal in the atmosphere.

Where individual vents are connected to fixture drains which discharge directly into drainage stacks below the level of the base connection of an attendant vent stack, or which discharge directly into the horizontal drain adjacent to the base fitting of a drainage stack, the principal function of such individual vents is that of a vent stack. In such case, the size of the individual vent should be at least as large as is required for an attendant vent stack.

Individual vents connected to fixture drains which discharge directly into the sanitary building drain or horizontal branch thereof are subject to pneumatic effects prevailing in such lower piping of the system. As these effects may vary from one extreme to another depending upon the distance at which the fixture drain connection is made from points at which airflow is constricted in the building drain, the size and load on the drain, and the presence of other building drain connections which may be available for relief of pressure therein, the quantity rate of airflow to be conveyed by an individual vent in this instance may be determined only by analysis of the given installation.

Where individual vents are connected to fixture drains which discharge into zones of the drainage system wherein suds pressure conditions may occur, the size of such vents should be determined for suds flow rather than airflow unless a special suds relief vent is connected to the suds pressure zone. Appropriate sizes for suds flow may be computed by means of Eq. (8-4) and by reference to Table 8-2, for the quantity rates of flow corresponding to conditions in the particular piping arrangement.

The minimum size recommended for an individual vent is 1¼ in. This is based upon the fact that the most critical factor with regard to the capacity of a vent pipe is its diameter. Any decrease in diameter has a seriously adverse effect upon the capacity of a vent pipe. Allowance must be made for a slight amount of fouling at the vent connection to the fixture drain. Consequently, the minimum recommended individual vent size is the same as the minimum permissible fixture drain size, 1¼ in.

In the section "Standard Code Regulations" in this chapter, the same minimum size is stated for individual vents, but the size of the fixture drain is applied as the basis for sizing the individual vent. The required vent size is one-half that of the fixture drain, except that a 1½-in vent may be provided for a 4-in fixture drain of a water closet.

BRANCH VENT DROPS TO LOWER FLOORS

Rational analysis may be applied for the sizing of branch vent drops to lower floor fixtures. If each fixture drain of such fixtures is assumed to be installed vertically and to extend downward a sufficient distance so that flow reaches terminal velocity in the vertical drop, the fixture drain may be considered to have the same performance characteristics and airflow requirements as a drainage stack.

Also, if each individual vent is assumed to be connected at the top of the vertical drain, the vent may be considered to perform the same airflow function as a vent extension at the top of a vent stack, namely, that of supplying air into the drainage stack at the same rate as it is dragged down by the liquid flowing down the vertical drainage stack.

In drainage stacks of 1¼ in (31.3 mm) through 4 in (100 mm) diameter, terminal velocity is achieved in a drop of less than one normal story height. Ordinary airflow requirements for vertical fixture drains of that size range, and conveying capacity flow, are the same as for drainage stack vent extensions.

Consequently, a branch vent drop serving a number of individual vents may be considered to perform the same function as that of a vent header or combined vent serving as the single vent pipe extending above the roof for a number of vent extensions of soil and waste stacks and their attendant vent stacks. It follows logically then that the sizing of branch vent drops should be comparable with the sizing of vent headers or combined vents in certain respects.

In the sizing of vent headers, using the standard tables developed for sizing vent stacks, the column headed "Size of soil or waste stack" is disregarded. The size is based upon the sum of the fixture units of load of all drainage stacks vented through each section of the header. The developed length applied is that of the vent stack having the greatest developed length to the atmosphere above the roof.

In essence, this amounts to considering the vent header as a vent stack which serves an imaginary drainage stack having a load equal to that of the total number of fixture units of all the connected stacks. The vent header corresponds to an imaginary vent stack with a developed length equal to that of the actual vent stack having the greatest developed length of all connected to the vent header.

The logic of this basis for sizing vent headers is established on the fact that the total amount of flow in drainage piping is related to the number of fixture units of load connected to the drainage piping, rather than to the number and size of drainage stacks into which the various fixtures discharge. If a single stack was to be used in lieu of a number of drainage stacks, and to convey the same total load, the single stack almost always would have to be larger in size than any of the drainage stacks.

In any event, a single drainage stack could theoretically be used in lieu of a number of drainage stacks. However, in that case long horizontal fixture drainage branches would be required, resulting in excessive loss of headroom due to slope in long branch drains. These branches would also be subject to excessive incidence of stoppages.

The developed length of vent stack for a single drainage stack with numerous long horizontal offsets would be comparable in length with that of the most extreme run or traverse of the drainage stack. Hence, the basis cited for sizing vent headers may be deemed reasonable and rational.

The same basis may be applied to sizing branch vent drops. But, in this case, the vent stack connection for the branch vent drop may be considered to be the equivalent of the vent terminal of a vent header or combined vent. Atmospheric pressure may be considered to prevail in vent stacks as a normal condition during service, except for relatively short periods when peak flow occurs in drainage stacks.

During peak flow periods, positive or above-atmospheric pressure conditions occur in the lower sections of drainage stacks and their attendant vent stacks. This positive pressure condition in the vent stack simply assists flow in branch vent drops, since their principal function is to supply air to the various fixture drains served when flow occurs therein.

Consequently, the rational basis for sizing branch vent drops to lower floor fixtures is the same as for sizing vent headers or combined vents, except that the maximum developed length which should be applied for branch vent drops is that measured from the vent stack branch connection to the farthest fixture drain connection served (Fig. 8-1).

Using the standard tables developed for sizing vent stacks (Tables 8-5 and 8-6), note the vent pipe sizes suitable for the developed length applicable. Also note the number of fixture units of load suitable for each vent pipe size for the developed length applicable. Then, for that developed length, select appropriate vent pipe sizes for each section of the branch vent drop corresponding to the load in fixture units. This corresponds to sizing branch vent drops on a uniform pressure loss basis for their maximum developed lengths.

BRANCH VENTS

Branch vents which serve to connect more than one individual vent to a vent stack or to a vent terminal in the atmosphere should be sized in accordance with the principal function performed thereby in any given installation and the corresponding quantity rate of airflow required to be conveyed. Just as for individual vents, proper sizes may be established

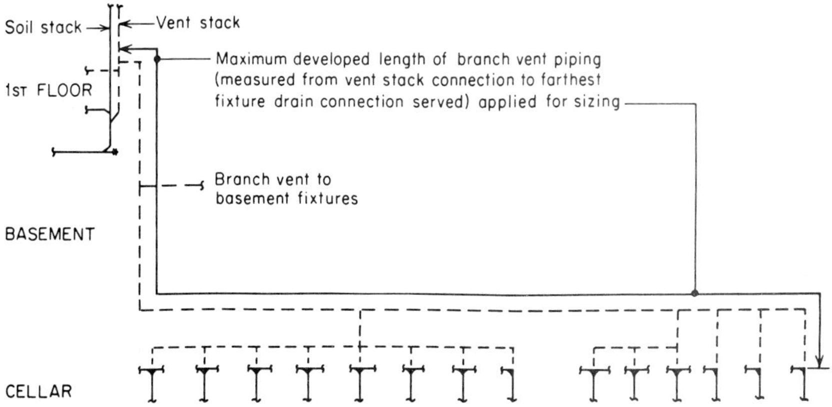

Fig. 8-1 Developed length applicable to sizing branch vent drops to lower floor fixtures.

by analysis of the arrangement of drainage and vent piping for any particular system.

Where several individual vents connect to a horizontal branch vent extending from a branch fitting in a vent stack and their respective fixture drains connect to a horizontal branch drain extending from a branch fitting in a drainage stack, a pressure relief circuit exists between the drainage stack branch fitting and the vent stack branch fitting. Each of the fixture drains and their individual vents form branch or parallel circuits for pressure relief and airflow between the drainage stack and vent stack. The proportion of the total airflow conveyed by each individual circuit is the same as would be the case in parallel circuits in a water supply system.

In all other respects the size of branch vents may be computed in the same manner as for individual vents, that is, based upon their principal function and the quantity rates of airflow or suds flow to be conveyed. A convenient basis for sizing branch vents is given in the section "Standard Code Regulations" in this chapter. The size in this case is based upon the number of fixture units connected to the branch vent and the developed length of the branch vent measured from its vent stack or stack vent connection to the farthest fixture drain connection served by the branch vent.

CIRCUIT AND LOOP VENTS

A circuit or loop vent is little more than an enlarged individual vent, connected to the fixture drain of the most upstream fixture discharging into a horizontal branch drain and provided to permit air to flow into the horizontal branch drain at a sufficient rate so as to prevent excessive aspiration effects from occurring therein. This is the principal function of a circuit or loop vent in most ordinary installations.

However, where a circuit or loop vent serves a horizontal branch drain which connects to a pressure zone of the sanitary drainage system, such as may occur adjacent to the base of a drainage stack or in the building drain, the principal function of the vent may be to prevent excessive back-pressure effects from existing in the horizontal branch drain. Thus, the quantity rate of airflow required to be conveyed by a circuit or loop vent must be determined in accordance with its principal function and corresponding airflow requirements, just as for individual or branch vents. For any given installation, the sizes of circuit or loop vents may be computed in accordance with their air and suds flow requirements in the same manner as was discussed for individual vents.

In the section "Standard Code Regulations" in this chapter, the size of circuit or loop vents is required to be at least one-half the diameter of the horizontal soil or waste branch to which they connect. This may be adequate for most ordinary installations requiring airflow to be conveyed to prevent excessive aspiration effects in the horizontal branch drain.

RELIEF AND YOKE VENTS FOR SOIL AND WASTE STACKS

The function of a relief or yoke vent, provided in special locations in drainage stacks of extended height, is to permit air pressure to be relieved from the drainage stack directly to the vent stack by flow through a short relief or yoke vent of the same size as the vent stack. Wherever flow interference may occur in such drainage stacks, the relief or yoke vent functions as an additional elevated vent stack base connection above the point of flow interference.

Sizing in this case is related to the sizing of vent stacks and their base connections. Hence, the proper size for relief and yoke vents is the same as that of the vent stack required for the drainage stack.

SUDS PRESSURE RELIEF VENTS

Suds pressure relief vents, provided at special zones in the drainage system where suds pressure may develop, must be adequate in size to permit suds to flow out of the zone at a sufficient rate so that the pressure rise in the zone does not exceed the design criterion. Such special vents are necessary to prevent suds backup conditions at fixtures which discharge into such zones and are equipped with traps protected by vents which are inadequate to perform the function of relieving suds pressure.

An ordinary fixture trap with 1 in (25 mm) of residual trap seal depth in service has a maximum resistance of 2 in (50 mm) of water column when subjected to back pressure of suds. This resistance is equal to a column of suds 7.42 ft (2.26 m) in height, at which level suds backup through the fixture trap seal may occur.

A suds pressure relief vent need not be of extensive length, for it is only necessary to relieve suds pressure from the pressure zone to a nonpressure zone downstream from and at a lower elevation in the sanitary drainage and vent system. Hence, the principal factor involved in this case is the diameter of a relatively short vent required to relieve the quantity rate of suds flow in the pressure zone. A simple guide in this case is the quantity rate of airflow conveyed down a drainage stack into the suds pressure zone, which may be determined from rates shown in Table 8-4, and the outlet discharge rates for suds at a flow pressure of 1 in (25 mm) water head, which may be determined from the capacities for various sizes of vent outlets in Table 8-2. The maximum permissible length of suds pressure relief vents can be further computed by means of Eq. (8-4).

From these tables and by computations, it can be seen that the size of the suds relief vent in most cases is approximately three-fourths that of the drainage stack and its horizontal drain at the base of the stack. Thus, this may be taken as a recommended guide in sizing such special relief vents.

VENTS FOR BUILDING SEWAGE SUMPS AND RECEIVING TANKS

Building sewage sumps and receiving tanks, other than pneumatic ejectors, operate at atmospheric pressure and receive sewage from sanitary subbuilding drains under gravity flow conditions. The only airflow required in a vent pipe from such an airtight receptacle is that which is displaced by sewage in entering the receptacle and that which is required to enter the receptacle as the sewage is ejected by the sewage pump. Thus, the quantity rate of airflow required to be conveyed by the vent for a building sewage sump or receiving tank is the maximum rate at which sewage either may enter the tank or be pumped therefrom, whichever is the greater rate.

Based upon the quantity rate of airflow required to be conveyed and the developed length of vent piping for a given installation, a proper size may be computed for such vents by means of Eq. (8-3) and by reference to Table 8-2.

STANDARD CODE REGULATIONS

Individual Vents Individual vents shall be at least 1¼ in in diameter and no less than one-half the diameter of the fixture drain to which they connect, except that a 1½-in individual vent may be installed where a 4-in fixture drain is provided for a water closet or similar fixture.

Circuit and Loop Vents Circuit or loop vents shall be at least one-half the diameter of the horizontal soil or waste branch to which they connect.

Branch Vents Branch vents connecting more than one individual vent to a vent stack or stack vent shall be in accordance with Table 8-1. In determining the size of such piping, the column headed "Size of soil or waste stack, in inches" shall be disregarded; and the size shall be based upon the number of fixture units connected thereto, and the developed length of the branch vent measured from its vent stack or stack vent connection to the farthest fixture drain connection served by the branch vent.

Vents for Building Sewage Sumps and Receiving Tanks Vents for building sewage sumps and receiving tanks, other than pneumatic ejectors, shall be sized as branch vents.

Relief and Yoke Vents for Soil and Waste Stacks Relief and yoke vents for soil and waste stacks shall be not less than the size of the vent stacks to which they connect.

Vent Headers Sections of a vent header and its vent extension through the roof shall be in accordance with Table 8-1. In determining the size of such piping, the column headed, "Size of soil or waste stack, in inches," shall be disregarded; and the size shall be based upon the sum of the fixture unit loads of the stacks vented through such section of the header, and the developed length shall be that of the vent stack having the longest developed length to the open air.

Vent Stacks The size of vent stacks shall be determined from Table 8-1, based upon the size of soil or waste stacks served thereby, the number of fixture units connected to the soil or waste stack, and the developed length of the vent stack. Such developed length shall be that measured from the lowest connection of the vent stack with the soil stack, waste stack, or building drain, to the vent terminal in the open air.

Table 8-2
OUTLET DISCHARGE RATES FOR AIR AND SUDS
Flow Pressure = 1 in (25 mm) Water Head

Outlet diameter d_o		Air discharge q_d		Suds discharge q_d	
in	mm	gpm	L/s	gpm	L/s
1¼	31.3	171.2	10.8	39.5	2.49
1½	37.5	246.5	15.6	56.9	3.59
2	50.0	438.3	27.7	101	6.37
2½	62.5	684.8	43.2	158	9.97
3	75.0	986.1	62.2	228	14.4
4	100.0	1753	110.6	405	25.6
5	125.0	2739	172.8	633	39.9

FLOW OF AIR IN VENT PIPING

The maximum rate at which air may discharge from an outlet into the atmosphere in practice may be expressed as follows:

$$q_d = c_d q_i = c_d(2.448\ d_o^2\ V_i) = c_d(2.448\ d_o^2 \sqrt{2gh_m})$$
$$= c_d(19.65\ d_o^2 \sqrt{h_m})$$

where q_d = quantity rate of air discharge from outlet in practice, gpm
q_i = ideal rate of discharge from outlet, gpm
c_d = coefficient of discharge for outlet
d_o = diameter of outlet, in
V_i = ideal velocity, fps
g = gravitational acceleration, 32 ft/s²
h_m = head measured in vent pipe during flow, ft of air column

When the value of the coefficient of discharge c_d is assumed to be 0.67, a reasonable value for ordinary pipe outlets, the equation for the maximum rate of airflow which may be obtained from an outlet in practice has the following simplified form:

$$q_d = c_d(19.65)d_o^2\sqrt{h_m} = 0.67(19.65)d_o^2\sqrt{h_m}$$
$$= 13.17d_o^2\sqrt{h_m}$$

In vent piping, the design criterion calls for limiting the air pressure differential above or below atmospheric pressure to no more than 1 in (25 mm) of water column. In terms of feet of air column, this may be established based upon the density of dry air at 70°F (21.1°C) as follows:

$$h_a w_a = h_w w_w$$

$$h_a = \frac{h_w w_w}{w_a} = \frac{1 \text{ in} \times 62.408}{12 \text{ in} \times 0.07512} = 69.23$$ ft (21.1 m) of air column or head [the equivalent of 1 in (25 mm) of water column or head]

Table 8-3
COMPUTED AIRFLOW RATES REQUIRED FOR VENTING HORIZONTAL SANITARY DRAINS FLOWING HALF FULL

Diameter of drain		Slope of drain		Air and water flow rates	
in	mm	in/ft	mm/m	gpm	L/s
1¼	31.3	½	41.6	5.5	0.35
1½	37.5	½	41.6	8.3	0.52
2	50.0	¼	20.8	9.7	0.61
2½	62.5	¼	20.8	17.6	1.11
3	75.0	¼	20.8	28.6	1.80
4	100.0	¼	20.8	57.0	3.60
5	125.0	¼	20.8	96.5	6.09

This conversion factor may be applied to establish a simplified equation for the quantity rate of air discharge into the atmosphere from outlets, of various nominal sizes, in a vent pipe wherein the pressure is 1 in (25 mm) of water column, as follows:

$$q_d = 13.17 d_o^2 \sqrt{h_m} = 13.17 d_o^2 \sqrt{69.32} = 13.17 \times 8.32 d_o^2 = 109.57 d_o^2 \tag{8-1}$$

Suds discharge rates may be established in a similar manner. The feet of suds column or head equivalent to 1 in (25 mm) of water column may be calculated, based upon the assumption of 1.4 lb/ft³ (22.5 kg/m³) as the density of old or regenerated suds, as follows:

$$h_s w_s = h_w w_w$$

$$h_s = \frac{h_w w_w}{w_s} = \frac{1 \text{ in} \times 62.408}{12 \text{ in} \times 1.4}$$ = 3.71 ft (1.13 m) of suds column or head [the equivalent of 1 in (25 mm) of water column or head]

Applying this conversion factor, an equation may be established for the quantity rate of suds discharge into the atmosphere from outlets, of various nominal sizes, in a vent pipe wherein the pressure is 1 in (25 mm) of water column, as follows:

$$q_d = 13.17 d_o^2 \sqrt{h_m} = 13.17 d_o^2 \sqrt{3.71} = 13.17 \times 1.925 d_o^2 = 25.3 d_o^2 \tag{8-2}$$

Comparison of the equations for suds and air discharge rates indicates that the suds discharge rate is just 25.3/109.57, or 23.1 percent of the air discharge rate for the same flow pressure. Table 8-2 shows the quantity rates at which both air and suds may discharge into the atmosphere from various sizes of vent outlets, under a flow pressure equivalent to a 1 in (25 mm) water column or head.

PRESSURE LOSS DUE TO FRICTION IN PIPING

When air flows through vent piping to relieve air pressure resulting from the flow of liquids in sanitary drainage piping, a continuous loss of pressure occurs along the piping in the direction of flow. This pressure loss is due to friction generated between the moving air and the inner surface of the vent piping in view of the fact that turbulent flow conditions prevail at the maximum rates of flow which vent pipes are designed to convey.

The amount of pressure or head loss due to friction in this case is dependent upon such factors as physical properties of air, such as its density and temperature, roughness of the interior surfaces of the pipe; length of the pipe; diameter of the pipe; and the velocity at which air flows through the pipe.

Darcy's formula for pipe friction loss may be applied to calculate the head loss due to friction developed by airflow in vent piping. It is as follows:

$$h_f = \frac{fLV^2}{D2g}$$

where h_f = head loss due to friction, ft of air column
f = coefficient of friction corresponding to the roughness of the pipe surface and diameter of the pipe
L = length of piping, ft
D = diameter of piping, ft
V = velocity of flow, fps
g = gravitational acceleration, 32.2 ft/s²

PERMISSIBLE LENGTH OF VENT PIPING

For more convenient expression of pipe friction loss in, and permissible length of vent piping, the Darch pipe friction formula and the adapted flow formula may be combined and their terms converted to those more generally used in practice, as in the following:

$$h_f = \frac{fLV^2}{D2g} \qquad q = 2.448\, d^2 V$$

$$V = \frac{q}{2.448 d^2}$$

Hence,

$$h_f = \frac{fLq^2}{(d/12)(2.448)^2 d^4 (64.4)}$$

$$= \frac{fLq^2}{(32.16) d^5}$$

$$= \frac{(0.03109) fLq^2}{d^5}$$

and

$$L=\frac{h_f d^5}{(0.03109)fq^2}$$

where h_f = head loss due to friction, ft of fluid column
f = coefficient of friction corresponding to the roughness of the pipe surface and diameter of the pipe
L = length of piping, ft
D = diameter of piping, ft
d = diameter of piping, in
q = quantity rate of flow, gpm
g = gravitational acceleration, 32.2 ft/s²
V = velocity of flow, fps

The maximum permissible length of vent piping may be expressed in accordance with the design criterion for vent piping which requires that the seals of fixture traps be subjected to an air pressure differential of not more than 1 in (25 mm) of water column above or below atmospheric pressure. As the permissible head loss due to friction in this case is 1 in (25 mm) of water column, the equivalent head for air or suds may be applied in the preceding equation to determine the permissible length of vent piping for airflow or for suds flow.

For airflow, an equivalent head of 69.23 ft (21.1 m) of air column, based upon the densities of water and air at 70°F, may be applied. This results in the following formula for the maximum permissible length of vent piping for airflow:

$$L=\frac{(69.23)d^5}{(0.03109)fq^2}=\frac{2226d^5}{fq^2} \tag{8-3}$$

Similarly, for suds flow, an equivalent head of 3.71 ft (1.13 m) of suds column, based upon the densities of water at 70°F and that given for old or regenerated suds, may be applied. This results in the following formula for the maximum permissible length of vent piping for suds flow:

$$L=\frac{(3.71)d^5}{(0.03109)fq^2}=\frac{119.3d^5}{fq^2} \tag{8-4}$$

From these two equations, it can be seen that the permissible length of vent piping for suds flow is just 119.3/2226, or 5.36 percent as much as is permissible for airflow; and conversely, the permissible length for airflow is 2226/119.3, or 18.7 times as much as for suds flow.

Table 8-4
COMPUTED AIRFLOW CAPACITY REQUIRED BY ATTENDANT VENT STACKS

Diameter of drainage stack		Airflow rate for fraction of drainage stack area occupied by water at terminal velocity			
		$r_s = 6/24$		$r_s = 7/24$	
in	mm	gpm	L/s	gpm	L/s
1¼	31.3	15.0	0.95	15.8	1.06
1½	37.5	24.3	1.53	24.8	1.56
2	50.0	52.5	3.31	54.8	3.46
2½	62.5	95.4	6.02	99.8	6.30
3	75.0	156.0	9.84	164.0	10.3
4	100.0	333.0	21.0	348.0	22.0
5	125.0	606.0	38.2	633.0	39.9
6	150.0	1008.0	63.6	1065.0	67.2
8	200.0	2127.0	134.2	2220.0	140.0

Table 8-5
SIZE OF VENT STACKS FOR WET VENTING BATHROOM GROUPS

Number of wet-vented fixtures	Diameter of vent stack	
	in	mm
1–2 bathtubs or showers	2	50.0
3–5 bathtubs or showers	2½	63.5
6–9 bathtubs or showers	3	75.0
10–16 bathtubs or showers	4	100.0

Section **9**

Special Plumbing Systems

Routine design of plumbing systems is a complex task if the designer is to produce a system that is both efficient and economic. When special designs are required, the task becomes even more complex because data often are not available.

This section presents design techniques for a number of "offbeat" elements in a plumbing system. Items included in this section cover topics such as chilled-water drinking systems, distilled-water systems, demineralized-water systems, irrigation systems, swimming pool systems, decorative pools and fountains, waterfalls, water jets, compressed-air systems, vacuum air systems, etc.

Using the data provided here, the designer can prepare either a preliminary or final design for any of a variety of specialized plumbing systems. The data have been proven in numerous working applications wherein the design met the operating specifications.

1. Chilled-drinking-water systems.

a. Principles of design.

(1) The minimum number of drinking fountains (water coolers) required in a building or area of a building is usually governed by code. Most codes today also have requirements regarding drinking fountains for the handicapped.

(2) Types of drinking fountains include:

(a) Projecting type.
(b) Semirecessed type.
(c) Recessed type.
(d) Simulated recessed type.
(e) Unit water coolers.
(f) Cafeteria water coolers.

(3) Drinking fountains for the handicapped must be projecting type, with projection of at least 18 in and mounted 34 to 36 in above the floor.

(4) Provide all drinking water outlets with chilled water at 50°F.

(5) Where drinking fountains are stacked in the same location and the equipment space is reasonably adjacent, consider using central (or subcentral) circulating chilled-water systems as they will be economically feasible.

(6) In areas where drinking fountain locations are scattered or irregular, provide local chilled-water units.

(7) Build local chilled-water units either into the wall under the drinking fountain, exposed in a closet behind the drinking fountain, into the bottom of the semirecessed drinking fountain, into the back of the surface-mounted drinking fountain, or use unit water coolers, as per the architect's preference in each location. The units built into the wall or exposed in a closet may also feed drinking fountains

on the floor above and/or below in the same location.

(8) Whenever available, provide refrigerated drinking fountains (water coolers) with a precooler to save on energy costs of producing chilled water.

(9) Unit water coolers are sometimes provided with an integral hot-water unit and dispense-faucet to provide hot water for making instant coffee.

(10) Cafeteria water coolers are usually provided with push-back glass filling faucets. Sometimes they are also provided with a bubbler.

(11) All central and subcentral chilled-water systems should be circulating type with all bronze pumps.

(12) Where more than one riser is required, the circulation line can act as one of the feed risers.

(13) In up-feed risers, connection to the circulating line should be taken, where possible, at the ceiling below the top branch.

b. Equipment.

(1) All local chilled-water units should be air-cooled.

(2) The central chilled-water units should be air-cooled (whenever possible), air-cooled with blower; air- and water-cooled, or water-cooled, as size, location, and the requirements of the local authorities dictate.

(3) The central units should be sized to provide the required 45°F water to serve the people using the system, plus compensation for the circulating piping heat gain and the pump heat input.

(a) Where the number of people involved cannot be determined, use 5 gph/drinking fountain.

(b) The capacity required for people served: Permanent population = 1 gph/20 people; places of assembly or transient population = 5 gph/drinking fountain.

(c) The heat gain in gph/100 ft of piping, assuming 45°F circulating water, 80°F inlet water, and a 90°F room, is: for 1/2 in, 2.0; for 3/4 in, 2.2; for 1 in, 2.3; and for 1 1/4 in, 2.4.

(d) The heat gain from the circulating pump in gph, assuming 45°F circulating water, 80°F inlet water, and a 90°F room, is: for 1/4 hp, 2.2; for 1/3 hp, 2.9; for 1/2 hp, 4.4; for 3/4 hp, 6.5; and for 1 hp, 8.7.

(e) Select the unit on the basis of the required gph (80°F to 45°F). The ratio of gph to storage is fixed by the manufacturer.

(f) Consult the manufacturer about the compressor for his unit, depending on the type of cooler selected.

(g) The circulating pump should be the gpm selected by the manufacturer for his unit, with the head a total of his required head plus the head of the reduced gpm circulated through the system.

(4) Base the required system circulation (gpm) on a minimum of 1 gpm or on the following gpm's/100 ft of circulated piping. (Use total actual footage of circulated piping for gpm calculation, but only the equivalent length of the longest run for the head calculation.)

Pipe size, in	gpm/100 ft[a]
1/2	0.29
3/4	0.30
1	0.33
1-1/4	0.37
1-1/2	0.40

[a]Based on limiting temperature rise to 5°F in a 90°F room.

(5) Size local cooler units, not an integral part of the dispensing unit, on the number of outlets (5 gph each) or, if known, the number of people (at 20 per gph) served, delivering water at 50°F.

(6) Provide filters on all central units. Provide a minimum of two (each 1/2 size) for each system (with as many as required for larger systems). Separately valve each filter.

(7) For developing equipment, consult the chilled-water unit manufacturer for his recommendation of the unit usually provided to supply a given piece of equipment.

  (a) Locate the unit as close as possible to the equipment.

  (b) Provide oversized heavy-duty filters, to handle the continuous-flow requirements.

2. Distilled-water systems.

a. Principles of design.

(1) Provide distilled water in laboratories and/or at corridor dispensing stations in the laboratory areas, in pharmacies, and elsewhere, as required by the program. Review with the owner the expected quality of water (purity).

(2) Where requirements are numerous in an area, provide a central system.

(3) Where requirements are widely isolated, provide local units.

(4) Locate central units in the penthouse and feed by gravity to outlets.

b. Piping.

(1) Consult the owner regarding the piping materials to be used, because some owners have very strong feelings about the matter. Possible materials are:

(a) Block tin-lined brass pipe (ID 1/8 in less due to lining).
(b) Block tin-lined copper tubing (ID 1/8 in less due to lining).
(c) Polypropylene pipe.
(d) Glass pipe.
(e) PVC pipe.
(f) Type 304-L or 316-L stainless-steel tubing.
(g) Aluminum pipe (Type 3003).

(2) Avoid glass pipe, if possible, because it is fragile and should not be used when glass washers with solenoid valves are involved because it will not stand up under shock conditions.

c. Equipment.

(1) Central units should be duplex (each one half size) hospital-type, steam-heated completely automatic stills, and storage tank.
(2) Size the tank at five to ten times the hourly capacity of the stills.
(3) The following are the available tank materials and sizes:

(a) Tin-lined copper (round vertical), 5 to 300 gal.
(b) Tin-lined (steel or aluminum) rectangular, 100 to 1,500 gal.
(c) Type 304 stainless-steel (round vertical) 10 to 1,000 gal, (round horizontal) 300 to 3,000 gal.

(4) Lacking detailed requirements from the owner, size stills on the basis of two people per 20 x 10 ft module, allow 1-1/2 gpd per person plus 25 percent wastage, plus 50 percent for growth.

(a) Allow 1-1/2 gph per corridor outlet.
(b) Select the stills to generate daily requirements in 20 h.

(c) For glass washers, allow 45 gph (total 160 gpd).

(5) Note that stills can be operated on water pressures as low as 10 psi, and are not restricted to the 40 psi stated in their catalog. However, where less than 40 psi is used, automatic valves on the still must be specified as air-operated; and an air connection must be obtained from HVAC control air system. In all cases, the operating water pressure at the still should be stated in the specifications.

3. Demineralized (Deionized) water systems.

a. Principles of design.

(1) Provide demineralized water in laboratories and/or at corridor dispensing stations in the laboratory areas, and elsewhere, as required by the program. Review with owner the exact expected quality of water (purity).
(2) Where the requirements are numerous in an area, provide a central system.
(3) Where the requirements are few or isolated, provide local cartridge-type units.
(4) Locate the demineralizers in the basement with a pressure feed to the various outlets.

b. Piping.

(1) Pipe materials: See Section 3.

c. Equipment.

(1) The demineralizers should be completely automatic two-bed ion-exchange units with all required accessories, all selected for use on the raw water available and to produce water conforming to the owner's requirements.
(2) Factors for sizing demineralizers:

(a) Allow 0.5 to 1 gpd per person for classroom laboratories (assume two classes per day).
(b) Allow 1 to 1.5 gpd per person for other laboratories.
(c) Allow 5 gpd for classroom preparation room outlets.
(d) Plus wastage factor of 25 percent.
(e) If feeding stills, add that load.
(f) If feeding glass washers, add that load.
(g) Allow 25 gpd for pipette washer.

(3) With the chemical content of the water and the gpd load known, the time between regenerations can be calculated. Tentatively, select equipment with between a minimum of a 7 day run between regenerations and an optimum of 2 weeks. Next, calculate the average peak flow rates of the system. If the tentatively selected equipment or equipment slightly larger can handle the expected flow rates, that is the selection. However, if the equipment to handle the flow rates is considerably larger, use tentative selected equipment plus a storage tank.
(4) If raw-water booster pumps are required, they can be normal IBBM units. If deionized water booster pumps are required, they must be all stainless steel (required with tank systems). The water-distribution piping should be recirculated back to the tank by the booster pumps through a back pressure valve.
(5) Tanks: See Section 3.
(6) Depending on the raw-water conditions, a carbon filter and/or a softner may be required ahead of the deionizer.

4. Reverse osmosis (RO) water systems.

a. Principles of design.

(1) Provide RO water in laboratories and/or

corridor dispensing stations in the laboratory areas, and elsewhere, as required by the program. Review with owner the exact expected quality of water (purity).

(2) Locate the units in the basement with a pressure feed to the outlets.

b. Piping.

(1) Pipe materials: See Section 3.

c. Equipment.

(1) As the systems are specialized to the project requirements, after obtaining a complete raw-water analysis and the owner's requirements, consult with the manufacturer regarding equipment selection.

5. Water treatment.

a. Principles of design.

(1) Provide water treatment, where necessary, to furnish a piece of equipment with water of the quality essential for its operation.

(2) Water treatment equipment should be of a type and size as required for the duty that it has to perform.

b. Piping.

(1) Limit copper and brass piping carrying completely softened water to a maximum velocity of 5 fps.

c. Equipment.

(1) Water softeners.

(a) Resin exchange units (usually with automatic regeneration) complete with brine tanks, interconnecting piping, and all required accessories and controls.

(b) Normally units are sized on the gpm flow requirements of the system or equipment served. However, where gpm requirements are high compared with gph requirements (as for laundry washers), it may be more practical to base the size of the unit on the gph requirements (average gpm) and provide a storage tank to supply the peak gpm flows.

(c) Hardness is expressed in grains per gallon as calcium carbonate equivalents.

(d) Normal resins have a softening capacity of 30,000 grains/ft^3 when regenerated with 15 lb of salt per cubic foot of resin.

(e) Softening capacity between regeneration equals the cubic foot of resin x 30,000 divided by the grains per gallon of the water softened.

(f) Acceptable periods between regenerations depend on the operating schedule of system or equipment being served. Minimum should be one day.

(g) If 100 percent continuous supply is required, multiple units must be provided to maintain supply during regeneration of one of the units.

(h) Automatic regeneration can be actuated by the quantity of water used or by a time clock.

(2) Filters.

(a) The type of filter used depends on the filtration job it has to do and the required quality of the effluent.

(b) Filters may be sand, diatomite, cartridge, or carbon type as required by (a) above.

6. Nonpotable-water systems.

   a. Principles of design.

      (1) Where required by code, completely separate by an air gap the water piping serving flushing rim floor drains, bidets, mortuary equipment, autopsy equipment and other code required equipment from the other building water piping.
      (2) Use an over-the-rim supplied surge tank above, with gravity feed; or where this is not possible, use an over-the-rim supplied surge tank with a booster pump or a hydropneumatic pressure system to accomplish this.

   b. Piping.

      (1) Piping materials: See Section 3.

   c. Equipment.

      (1) Equipment: See Section 3.

A. Irrigation systems.

1. Principles of design.

   a. Provide irrigation systems where requested by the owner and/or the architect.
   b. Determine from the architect the exact areas involved, and the type of planting, i.e., grass, flowers, shrubbery, trees, etc., in each area.
   c. Provide the water supply to irrigation systems with approved backflow preventers. Check the code for detailed local requirements.
   d. Various types of plant materials require varying amounts of water and different types of heads; consequently a single zone of sprinklers should not be used to accommodate two different types of plant material (e.g., turf and shrub beds). Select the proper type of head for the growth it is to serve, as recommended by the head manufacturer.

e. Types of heads available.

(1) Spray type: Pop-up or stationary. Best for lawns. Operates on relatively low pressure. Triangular spacing preferred, using 60 percent of diameter of throw (generally up to 15 ft radius).
(2) Bed spray heads: For shrubs, flower beds, and ground cover. Installed on risers either above or below the foliage.
(3) Strip or line type: Stationary for narrow areas and edges.
(4) Bubbler heads: For soaking small areas. Installed above ground. Flood type for very small areas and stream type for slightly larger areas, and under extremely dense foliage gives low water rate.
(5) Rotary type: For large areas, has long throw, low water rate, and requires higher pressures. Good for steep slopes. Hard to control in wind. Square pattern preferred using 60 percent of diameter of throw.

    (a) Gear drive: Pop-up. Small 16 to 30 ft radius, and large 30 to 80 ft radius. Valve in head type over 80 ft radius.
    (b) Impact drive: Installed above ground or pop-up over 80 ft radius.
    (c) Cam drive: Pop-up. Troublesome if water is not clean.

(6) For athletic fields, provide special rubber covered heads.
(7) Risers for above ground heads should be adjustable type.

f. Uses of various type heads.

(1) Spray heads (pop-up or stationary) are generally used in residential or small areas where large diameters are not required or where water quantities are limited. Spray coverage is excellent and uniform.

(2) Rotary heads.

(a) Gear drive (pop-up), medium-sized installation requiring greater radius than 15 ft.

(b) Impact type heads (pop-up or fixed) are used for large installations such as large commercial sites, golf courses, or cemeteries. These heads have large areas of coverage and use high quantities of water at good pressures. Impact heads do not give as even a spray coverage as other types of heads.

(3) Select special type heads for trees, shrubbery, strips, edges, etc.

(4) Try to use as many 360° heads as possible in a given system since the costs for installation of full-circle and part-circle heads are approximately the same while the area covered is much less. The use of an excessive number of part-circle heads will thus result in a higher unit price per area sprinkled.

(5) If part-circle heads are used, check to see if the watering density is different from the full-circle heads. If there is a substantial difference, the part-circle heads should be valved on a separate zone to permit the varying of the time cycle to avoid overwatering or underwatering. Varying densities with part-circle heads may occur with any type of spray head.

(6) Overlap of sprinkler patterns is required. Layouts of sprinklers where spray patterns are tangent are incorrect. In laying out the system, the manufacturer's recommendations for spacing are to be followed; not the criteria for spray radius. Radius criteria is not to be used, since this figure refers to effective radius in some catalogs and extreme radius in others.

(7) In areas which are confined by sidewalks, buildings, parking lots, etc., provide 180° heads at the edge of these areas which throw into the area to be irrigated. Do not locate heads which could wet areas not intended to be watered.

(8) In areas where wind is anticipated as a problem, the following corrections should be made:

   (a) Where wind is in a constant direction, shift sprinklers into the wind by 1/2 to 1 ft per mph of wind velocity. (Use average wind velocities during the watering season, not extreme values.)
   (b) Where wind direction changes, reduce spacing (on triangular basis) by about 10 percent per 10 mph of wind velocity.
   (c) For part-circle heads, increase or decrease the part-circular coverage as required by examination to give even coverage.
   (d) For heads along walks, roads, etc., where wind is a factor, use 200° heads instead of 180° heads when the wind is blowing away from walks and 140° heads instead of 180° heads when the wind is blowing toward walks.
   (e) On shallow slopes or in valleys, use the actual surface dimensions and area for calculation, not the plan dimensions (compensate for slope). Heads may be tilted slightly to give optimum spray coverage. However, this tilting and adjustment should be made in the field. On steep slopes, tilt head slightly and cut back the upstream side slope. Spacing on steep slopes should be approximately the same (or perhaps a bit tighter) as for normal areas.

g. Sections should be fed as near the center as practical to minimize pipe sizes.
h. Automatic sectional valves should be selected with or without pressure regulators as required.

    (1) Valves should be installed inside the building when ever practical.
    (2) Underground valves should be provided with plastic valve boxes.
    (3) All automatic valves should be preceded by control valves.
    (4) Automatic valves should be carefully selected for the required flow, often smaller than line size.

i. When manual sectional valves are used, they should be located so that the operator will not be wet by the spray, and they should be easily located.
j. Each section (zone) should be pitched to and provided with an automatic drain valve.
k. Minimum cover on piping should be 18 in.
l. Heads in areas subject to heavy equipment travel over them should have swing-type connections.
m. Irrigation systems should be provided with automatic controllers to properly time the watering of each section and to stagger the operation of the various sections to minimize the required peak water demand (flow). Controllers should be the heavy-duty type.
n. Where available pressure is insufficient for the requirements of a properly designed system, provide a booster pump. Booster pumps should be sized on the estimated peak flow required and the pressure differential required to obtain the required pressure at the farthest head.
o. As watering is usually done in off hours, its demand normally is not added to the peak domestic demand on the project.

2. Pipe sizing.

    a. Size piping based on the estimated peak flow in that particular piece of pipe, giving due consideration to the sectionalizing of the whole system.

b. Calculate friction losses based on the appropriate friction tables for the type of pipe being specified. Where an option of pipe materials is specified, use the friction table of the acceptable pipe with the highest friction losses.

3. Piping.

   a. Piping inside the building should be as for domestic water supply systems.
   b. Underground piping should be cast-iron, ductile-iron, asbestos cement (Transite), or plastic of proper pressure rating: polyvinyl chloride (PVC) or polypropylene (PP) with socket welded fittings, or polyethylene (PE) with banded insert fittings. Except for large volume systems (large pipe sizes), generally plastic will be the pipe material used.
   c. Plastic piping can be specified using the SDR pressure designations. This allows specifying a greater variety of pressure ranges depending on the specific job conditions and enables incorporation of economies in the system.
   d. Control valves on plastic piping should be plastic and provided with plastic extension boxes to grade.
   e. Where automatic control valves are located out in the system rather than in the building, pressure rating of plastic piping ahead of valves should be selected generously to allow a safety factor for water hammer.

B. Swimming pools.

1. Principles of design.

   a. Swimming pools, except for private residential pools (and in some localities, private pools) are governed by code.

      (1) Usually the Department of Health code, rather than the plumbing code (New York City-both) is followed.

(2) Design swimming pools in accordance with the requirements of all local authorities and the National Swimming Pool Institute, whichever is the more restrictive.

(3) The requirements usually govern the following:

   (a) Time of turnover (recirculating rate); usually 6 to 8 h.
   (b) Types of filters and filtration rates.
   (c) Spacing of main drains and grate area requirements.
   (d) Spacing of inlets.
   (e) Spacing and capacity of gutter drains and surge tank or skimmers (usually outdoor pools only).
   (f) Heating requirements.
   (g) Chlorination requirements.
   (h) Draining requirements (both time and to which sewer system).
   (i) Fill and makeup requirements.

b. Consult your supervisor about the design of all swimming pools.

c. Types of pools are as follows:

   (1) Wading pool: maximum water depth 2 ft.
   (2) Swimming pool: depth over 2 ft.
   (3) Hydrotherapeutic pool: a swimming pool (usually approximately 95°F) in a hospital used for treating patients.

d. Where no code exists or an item is not covered in the code, use the following design criteria:

   (1) Recirculation rate.

      (a) Swimming pools; 6 to 8 h turnover. Lightly loaded pools can use 8 h, but heavily loaded pools should use 6 h.
      (b) Hydrotherapeutic pools should use 4 to 6 h.
      (c) Wading pools should use 4 h.

e. Main drains should be sized for 100 percent of the recirculation rate with grates having an area of at least 4 times the area of the outlet pipe or an area that produces a maximum velocity of 1-1/2 ft/s under maximum flow conditions. The drains should be located a maximum of 15 ft from the side wall with a maximum of 30 ft between drains.

f. Gutter drainage system should be sized for 50 to 100 percent of the recirculating rate with a minimum 2 in drain outlet. Drains should be spaced a maximum of 15 ft apart. Grate area should be at least 1-1/2 times the area of the outlet pipe. Special patented continuous-flow gutters (with adjusting weirs) and supply assemblies are sometimes requested. Check with local code authorities about its acceptability.

g. If used instead of gutters, skimmers should be provided for every 400 to 500 ft^2 of pool area and sized for 80 to 100 percent of the recirculation rate with a minimum of 30 gpm per skimmer, or if over 8 in wide, 3.75 gpm per inch of weir length. Provide skimmers with equalizer pipes. Some model skimmers also include an integral vacuum cleaning connection, thus eliminating the need for separate vacuum cleaning connections on the pool wall.

h. Provide surge (balancing) tanks for pools (except private pools) using sand filters, unless skimmers or patented continuous-flow gutters are used. Allow surge capacity of 1/2 to 1 gal/ft^2 of pool area. Surge tanks are not usually required with vacuum diatomite filters, skimmers, or patented continuous-flow gutters.

i. Locate inlets around the perimeter of the pool to give an even distribution. Maximum spacing 20 ft and with an inlet within 5 ft of a corner.

   (1) Inlets should have an integral throttling device to balance the system.

   (2) Sometimes pools (mainly those used primarily for diving) are designed with their inlets in the bottom, using the gutter (or skimmers) for recirculation, and the main

drain only for emptying. If underground, this has the bad feature of placing the supply piping under the pool where it is very costly to get at for repairs.

j. Vacuum cleaning outlets should be spaced as required to cover the pool using not over 50 ft of hose. The size and type of cleaning tool should be selected based on the size and shape of the pool.
k. Chemical treatment equipment, hypochlorinators, should maintain a free chlorine residual of 0.6 to 1 ppm with a pH of 7.2 to 7.8.
l. Pool water temperatures should be from 76°F to 78°F.
m. Fill and makeup connections should have an air gap, either over the rim of the pool, the surge tank, the vacuum diatomite filter, or the makeup tank unless the code permits the use of an approved reduced-pressure back-flow preventer.

   (1) Makeup connections should be provided with a solenoid valve or solenoid pilot controlled valve, interlocked with the high-water alarm and the pump to prevent overflow or flooding if pump is not running.
   (2) When makeup water is introduced through an air gap, provide an open- and closed-type (modulating type is not necessary) float valve with speed controls.
   (3) Provide a full-size valved line for filling and a smaller-size bypass including float valve and throttling valve for makeup.

n. Provide all pool recirculating systems with an approved rate-of-flow indicator. New York City also requires that it be a recording chart unit.
o. Provide pressure gauges in the discharge of all recirculating pumps, the suction of all recirculating pumps, and the inlets of all strainers. Check whether the pump suction and strainer inlet gauge should be compound type.

p. Provide pools with drains in the deck surrounding the pool. Pipe these deck drains to the sanitary system for inside pools and the storm system for outdoor pools.

q. Provide underground pools (except small residential) with a tunnel around them to carry the piping and make it accessible, rather than burying the piping.

r. Provide underground pools, dipping into ground water, with hydrostatic relief valves in the bottom.

s. Carefully select all pool fittings for their purpose and location.

   (1) In selection, special attention must be paid to their proper connection to the architect's waterproofing system. Membrane requires lead flashing "burned" to the body (bronze) (not clamped) and elastomeric and liquid water-proofing requires a wide flange.

   (2) Drains must be without weep holes. If the drain is selected from the swimming pool section of the catalog, it has no weep holes. If it is selected from the roof drain or floor drains sections, it must be noted to be without weep holes. Carefully check this point on shop drawings.

t. Supply piping should have a maximum velocity of 8 fps. Recirculating return piping to pumps should have a maximum velocity of 5 to 6 fps.

u. When an aluminum pool is involved, all possible copper or copper alloys must be eliminated from the system.

   (1) In addition, an easily replaceable sacrificial piece of aluminum piping at least 2 ft long and with at least one elbow (the more the better) and having flanged or union ends should be provided in every line carrying water to the pool.

   (2) Provide insulating bushings, couplings, unions or flanges whenever any other metal contacts the aluminum.

2. Piping.

a. Fill and makeup water piping shall be as specified for domestic water piping.

b. As recirculated pool water is more corrosive than domestic water, it requires piping materials as good or better than that required for the domestic water.

(1) Inside the building, small piping should be copper tubing or red brass piping. Larger sizes should be bitumastic enamel (AWWA specifications), epoxy, or plastic-lined steel pipe and malleable or cast-iron fittings.

(2) Underground piping may be polyvinyl chloride (PVC) or polypropylene (PP) pipe and fittings.

(a) In small sizes it may be copper tubing.

(b) In larger sizes it may be cementlined cast-iron or ductile-iron bell and spigot pressure pipe and fittings or asbestos cement (Transite) pressure pipe with cast-iron or ductile-iron fittings.

(3) As most control valves are used for throttling, they should be ball valves in the smaller sizes and butterfly valves in the larger sizes.

c. Inside the building, plastic pipe and fittings may be used, if approved by your supervisor. Where piping is insulated, fire and smoke ratings of the insulation supercedes those of the pipe and fittings.

3. Equipment.

a. Filters.

(1) Rapid sand filters (pressure type): Rate 3 gpm/ft^2. Require at least three units as they must be backwashed at 3 to 4

times the recirculation rate. They are wasteful of water and require the most space. Require alum and soda ash feeders to operate.

(2) Vacuum diatomite filters: Rate 2 gpm/ ft^2. Require little water for backwash and minimum space. Due to fiberglass tank, they require no painting. The open tank can serve as the required air break and eliminate the need for a makeup, surge, or balancing tank.

(a) They require a body coat feeder.
(b) They require a pit to collect the used diatomite for manual removal. This is their major disadvantage.
(c) Precoating and backwashing can be automated but is expensive.
(d) Where a vacuum-type filter is used, it is imperative that an accurate section be drawn showing the elevations of the pool, the filter, and the filter room, because these vertical distances are critical in the selection of a filter and its control and must be made known to all manufacturers bidding on the job. In making these sections, consideration should be given to the foundation under the filter, as some filters require certain minimum height foundations due to their construction.

(3) High rate sand filters: Rate, 15 gpm/ ft^2. Backwashed at recirculating rate.

(a) The latest type therefore not yet recognized by some codes. May be allowed upon special request. So far, mainly used in the small sizes.

b. Pumps.

(1) The recirculating pumps should be flexible coupled horizontally split case, flexible

coupled, or close-coupled vertically split case, depending on the size. With sand filter systems, it is sometimes possible to use the recirculating pump as the vacuum cleaning pump by sliding it on its curve, if it is too large, or by providing two half-size recirculating pumps.

(2) A vacuum cleaning pump must be provided on sand filter systems where the recirculating pump (or pumps) cannot meet the vacuum system requirements, and on vacuum diatomite filter systems. Pump should be close-coupled vertically split case or flexible coupled horizontally split case, depending on the size.

(3) The head on the pump in a bottom recirculating system with a pressure filter and no surge tank should be the sum of the following: The pressure required at the outlet (for patented continuous-flow gutters allow 10 psi) coverted from psi to feet, the discharge friction, the loss through the heater, the loss through the filter, the suction friction, and the loss through the strainer.

(4) The head on the pump with a surge tank or vacuum filter should be the sum of the following: The pressure required at the outlet (for patented continuous-flow gutters allow 10 psi) converted from psi to feet, the lift from the pump to the pool water level, the discharge friction, the loss through the filter converted from psi to feet, the friction in the suction piping from the tank, and the loss through the strainer converted from psi to feet.

    (a) For vacuum diatomite filters, allow 20 ft for the loss through the filter.
    (b) For high-rate sand filters, allow 15 to 20 psi converted from psi to feet for the loss through the filter.

(5) NPSH requirements of pumps should be carefully checked.

(6) Pool pumps are manually operated.

(7) Pumps drawing from tanks should be provided with low-water cutoff switches.

   (a) Except for recirculating pumps pulling through diatomite filters, all recirculating pumps should be provided with full line size quick-opening basket strainers in their suction piping.

c. Heater.

(1) The heater should be a shell and tube heat exchanger. The heater must be able to heat the water on filling the pool and heat the required makeup. As these two requirements are widely different, the sizing must be a compromise, requiring more than one turnover to bring the freshly filled pool up to temperature, but not too long.

(2) In small sizes, usually all the water is passed through the heater with the required temperature rise.

(3) In large pools, it is usually more economical to pass only part of the water through the heater at an increased temperature rise, then blend it back with the bypassed water.

(4) Advise the HVAC project engineer of the steam requirements.

d. Sterilization equipment.

(1) Sterilization of the pool water to kill bacteria, and in addition for outdoor pools, to eliminate algae, is accomplished by the use of a hypochlorinator (chemical feeder).

   (a) Hypochlorinators in public pools are usually required to be provided in duplex.

   (b) This equipment may be automated.

(2) Automation requires an additional chemical feeder for pH control.
(3) The automation equipment gives direct reading of the chlorine residual and the pH of the pool water. It can also be bought to produce continuous reading charts, if the owner so desires.
(4) Hypochlorinators should be small, adjustable, positive displacement pumps (chemical feeders), built specially for the service, drawing sodium hypochlorite from a container and injecting it into the filter discharge piping to the pool after the last (hot-water heater) connection.
(5) Hypochlorinators must be provided with plastic containers to hold the required sodium hypochlorite, and the acid and/or alkali required for pH control.

   (a) Check that the units selected have a discharge pressure capability greater than that in the piping that they connect to.
   (b) In the table are suggested minimum sized hypochlorinators:

System volume, gal	Suggested size, gph[a]
10,000	0.14
18,000	0.25
42,000	0.58
60,000	0.83
72,000	1.0
114,000	1.6
180,000	2.5
360,000	5.
720,000	10.
972,000	13.5
1,300,000	18.

[a]Capacity at maximum adjustment rating.

(6) Hypochlorinators should be interlocked with the filter recirculating pump so that

they operate only when the pump is running.

e. In the table are the requirements of the various sizes of vacuum cleaning tools:

Size	Hose, in	gpm
12-in hand	1 1/4	35
15-in hand	1 1/2	50-60
18-in hand	2	70-80
24-in tow	2	80-100
30-in tow	2	100-120

(1) Use tow model in pools over 30 ft wide.
(2) Allow 32 ft of pump head for the tool plus friction in the hose, piping, strainer, etc.

C. Decorative pools and fountains.

1. Principles of design.

a. Determine from the architect exactly what is wanted in the way of a display, with respect to

(1) Number of nozzles and height of sprays
(2) Type of nozzles
(3) Special effects such as water falls
(4) Winterization
(5) Wind control

b. In conversations with the architect, point out the following: Vertical nozzles on a normal (not excessively windy day) can wet an area 1 to 1-1/2 times the spray height. Usually architects will not believe this. If you cannot convince them, design the system as they desire, and it will be throttled down to usable conditions after installation.

c. Water falls splash when they hit the water surface in the base pool; therefore, an adequate space must be allowed in the pool in front of a water fall (20-ft water fall requires at least 8 ft).

d. Generally decorative display fountains and pools are not covered by code (New York City does have code requirements) except for fill and makeup water connections.

   (1) Your supervisor should be consulted on the design of all display fountains and reflecting pools.
   (2) Decorative display fountains and pools should be designed on the basis of good engineering (hydraulics).

e. Except for very small fountains, all decorative fountains and pools should have a filtered recirculating system independent of the nozzle recirculating systems, because it must operate 24 h a day while the nozzles do not. The turnover rate should be approximately 12 h. In a winterized installation, this may have to be increased to keep temperature of the return water from being too high, to prevent steaming of the surface.
f. Because the architect is often not sure about what he wants to see, the pump and nozzles should be selected to allow for some upward adjustment after installation and the architect actually sees what it looks like. This is not necessary if the architect's height requirements are greater than is practical to operate the fountain.
g. A fountain is never fully designed in the design stage; the design is finally completed by adjustments in the field after installation.
h. In an unusual or complicated setup, it may be necessary to have a mockup or witness a test at the manufacturer's plant before completing contract documents.
i. To allow for future adjustments, an adequate number of throttling valves and rate-of-flow indicators should be provided, as well as a valved bypass between all jet recirculating pump discharges and suctions.

j. The pool water depth must be deep enough to allow for the proper installation of the size and type of nozzles selected.

   (1) Aerating jets require greater water depths than straight jets.
   (2) In New York City, the maximum allowable water depth is 18 in.

k. Before finalizing the pool water depths with the architect, consult with the electrical project engineer, because his lights may require minimum water depth that may govern in some cases. To save on weight of water in the pool, sometimes only the area immediately around the cluster of nozzles and lights is depressed to give the required water depth without increasing the depth of the entire pool.

l. In considering the free-board volume of a gutter or lower pool receiving the flow from a waterfall, remember that when the installation is operating, there is a measurable layer of water above the waterfall weir over the entire pool, which, when the pump stops, must be stored in the gutter or lower pool until the pump starts again. If space permits, it is also desirable to include provision for some storage of rain water to save on makeup water costs.

m. Nozzles are usually bronze and are available in many different sizes and shapes to give a wide variation of display patterns. However, all nozzles are basically one of two types: straight jets or aerating jets.

   (1) Straight jets, which give a solid stream of water.

      (a) They are best for long throws and the only type available for flat or fan stream or adjustable mushroom displays.
      (b) They are the only type that will operate completely out of water.
      (c) Mushroom nozzles may be obtained with a vertical jet in the center.

This type of nozzle has two inlet connections each connected to a separate system of piping due to the difference in pressure required for each display.

(2) Aerating jets, which pass the pumped water through a throat sucking in air and sucking up additional water from the pool.

(a) Since they do not require the pumping of all the water thrown into the air, the pumps and piping are smaller; however, due to the higher pressures required for the air and water pickup, horsepower may be greater.
(b) Depending on the model selected, they can pick up 2 to 4 times their pumped flow from the pool.
(c) Aerating jets produce a bushy foamy column that sparkles better under the lights than does the solid stream of a straight jet.

(3) Certain nozzles of the aerating-jet type are water-level-dependent and require close plus and minus water-level tolerances for proper operation.

(a) Pools with waterfalls from them automatically take care of this problem. However, the nozzles must be installed with the tops related to the water level when in operation (weir height plus the thickness of the sheet of water going over the weir).
(b) In other designs, automatic water-level controls (controlling makeup) are essential, rather than the manual makeup usually provided.
(c) Some models of aerating jets are adjustable, allowing a variation of the amounts of water and entrained air, which changes its appearance.
(d) Some models of aerating jets require

suction screens around their pool suction inlets.

(4) Check the manufacturer's catalog for recommendations on the need for a splash screen with the specific type of nozzle selected.

(5) To achieve the best possible appearance, most nozzles require linear, nontwisting flow into the nozzle, which means that there should be a straight section of pipe between the nozzle and a fitting and/or valve ahead of it. Where it is impossible to install a sufficient length of straight pipe ahead of the nozzle, flow vanes should be installed in the pipe ahead of the nozzle or results produced by the nozzle, as installed, accepted. Check with the manufacturer for recommendations when the above is not possible.

(6) For nozzles installed other than vertical, see Figure 9-1 for conversion to vertical height for selecting capacity requirements.

(7) All nozzles should be provided with ball joints beneath them to ensure that the rising water is absolutely vertical regardless of the workmanship.

n. Waterfalls, another type of water display, use weirs instead of nozzles to control the flow.

(1) Where the waterfall starts the display, a trough of water must be provided behind the weir to produce a constant even flow of water approaching the weir.

(2) Weirs should be provided with adjustable vertical bronze or stainless-steel weir plates to allow for even level adjustments, because it is impossible to construct masonery absolutely level.

(3) Following are some suggested flow rates over weirs. These are subject to modification to suit a particular project and the architect's desires.

Clearstream jet

Spray design calculations

Design Factors

A	E	C percent	D percent
5°	0.90	6	36
15°	1.33	11	46
25°	1.83	17	49
35°	1.94	22	51
45°	2.10	27	52
55°	1.80	36	53
65°	1.50	50	56
75°	0.90	99	59
85°	0.40	245	64

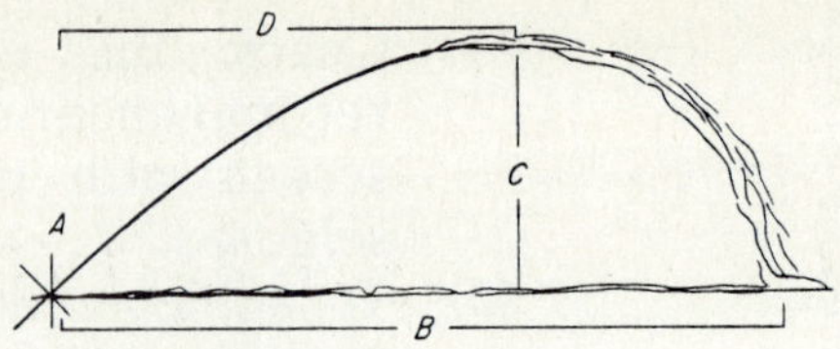

A = angle of nozzle elevation
90° 45° 0°
B = horizontal distance of throw from nozzle.
C = height of trajectory in percentage of *B*.
D = highest point of trajectory in horizontal distance from nozzle measured in percentage of *B*
E = multiplying or dividing factor for spray calculation
X = vertical spray height

To find:	How:
1. Horizontal distance of throw for a desired angle of spray when only the vertical spray-height is known.	1. Establish vertical spray-height factor X and multiply the same by factor E to achieve horizontal spray distance.
2. Performance requirements of spray-patterns other than vertical or the equivalent vertical spray-height performance requirements.	2. Establish horizontal distance of throw from nozzle and divide by factor E. This will give vertical spray-height, which is used to find performance requirements.
3. Trajectory of stream or spray.	3. Establish B, then calculate factors D and C thereof; combine the results with B to lay out the trajectory.
4. The jet elevation angle (factor A) for the specification of particular trajectories or spray effects.	4. Establish B, calculate factor C thereof, and read on the factors table the value of factor A as found on the same line as the result of the calculated factor C.

All spray design calculations are based on linear, non-turbulent flow or water into the jet having minimum directional adjustment and no wind. Where turbulence is present and best possible performance is desired, the use of flow-straightening vanes in the pipe leading to the jet, at a distance of at least ten times the diameter of the pipe from the jet, is recommended. Turbulence and twisting, nonlinear flow will cause early breakup and reduced spray-height and spray distance.

Fig. 9-1 Angle spray design factors. Information supplied by PEM Fountain Co., Richmond Hill, Canada.

(a) The higher the waterfall, the greater the flow must be to look good.

(b) With short weirs, generous flow rates can be used because the short footage does not give excessive total flow. However, with long weirs, flow rates must be conservative, because the long footage gives excessively high total flows, which can be costly in pipe size, pump size, and operating cost.

gpm/ft	Thickness of water over weir, in	Approximate normal maximum height of waterfall, ft
15	1/4 (0.3)	1
20	3/8 (0.4)	3
25	1/2 (0.5)	5
30	5/8 (0.7)	7
35	3/4 (0.8)	10
40	7/8 (0.9)	15
50	1 (1.0)	25
60	1-1/4 (1.3)	Over 25

o. Where nozzles requiring widely different pressures are involved, a decision must be made about the use of one recirculating system with pressure-reducing or throttling valves, or multirecirculating supply systems. (Return can be common from the same pool.) Generally, where the lower pressure is considerably lower or the quantity involved is much greater, multicirculating systems should be used to conserve energy.

p. Pool inlets in the filtration, chlorination, and heating recirculating system should be of sufficient number and carefully located around the pool and with respect to the drains, to ensure an even distribution of the chlorinated and heated water throughout the pool.

q. Pools are usually recirculated through the bottom drain. Occasionally due to neighborhood conditions (trees, shrubs, etc.), the architect or owner may request that recirculation be via a gutter or skimmers. Do not use skimmers where the water level is critical to the nozzle operation.

(1) Pools recirculated through the bottom drain are connected directly to the pump through a large strainer (no tank required).
(2) Pools recirculated through a gutter or skimmer require a tank in the equipment room to store sufficient water when the recirculating pump is shut down, to keep the pump supplied with water at startup until the first water pumped returns from the pool. In addition, tanks should be sized, if space is available, to store some rain water (free makeup water).
(3) Normal fill and makeup would be introduced into this tank, and must be automatically controlled at both pump-running and pump-off levels in the tank.

r. Recirculating outlets in the pool bottom should be provided with adjustable Anti-Vortex plates. This consists of a plate at least 4 in larger than the drain body, supported on four long threaded stud bolts brazed to the drain body.
s. Where a waterfall is involved and the nozzles do not supply sufficient water for a proper waterfall, additional water (a separate recirculating supply system) must be provided to the pool under water without breaking the surface.

(1) The inlet fitting for this is similar in type to the outlet drains described above, except the Anti-Vortex plate is called a Suppression plate.
(2) Inlet fittings to the trough feeding waterfalls should be provided with a Suppression plate, as described above.

t. All pools and fountains should be provided with overflows to prevent flooding in the event of abnormal rainfall or makeup valves stuck or inadvertently left open. For pools without tanks, overflows should be either freestanding (with dome) or wall type in the pool, as per architects desires and/or construction requirements.

u. All pool fittings should be carefully selected for their purpose and location.

   (1) In selection, special attention must be paid to their proper connection to the architect's waterproofing system. Membrane requires lead flashing "burned" (not clamped) to the body (bronze); and elastomeric and liquid waterproofing requires a wide flange.
   (2) Drains must be without weep holes. If the drain is selected from the swimming-pool section of the catalog, it has no weep holes; if selected from the roof-drain or floor-drain sections, it must be noted to be without weep holes. Carefully check this point on shop drawings.
   (3) For piping through waterproofing, use the waterproofing attached sleeve as a coupling in the piping.

v. If the pool is to be winterized, consult with the HVAC project engineer about recommended heating requirements.

   (1) Sometimes all heat is added in the filtration recirculating system and sometimes heat is also added in the jet system.
   (2) Give the project engineer your desired filtration rate (gpm). He may have to ask you to increase it if your rate would result in too high a temperature of the water returning to the pool (which would cause steaming on the surface of the pool on a cold day).

w. Normally all nozzles should be provided with individual throttling valves to even out the spray heights regardless of varying friction losses to the individual nozzles and for overall height adjustment.

x. When the nozzles are few and/or spread widely apart, the supply to each should come up through the bottom of the pool separately. Where the nozzles are closely spaced, provide a header or ring in the pool to feed the nozzles, with supply feeds as required through the bottom of the pool, to minimize penetrations of the waterproofing. Note that headers or rings in the pool usually increase the minimum required water depth.

y. Fill and makeup connections should have an air gap, unless the code permits the use of an approved reduced-pressure back-flow preventer.

   (1) Makeup connections should be provided with a solenoid valve or solenoid pilot controlled valve, interlocked with the jet pump to prevent overflow or flooding during nonoperating periods.
   (2) When makeup water is introduced through an air gap, provide an open- and closed-type (modulating type is not necessary) float valve equipped with speed controls.
   (3) Provide a full-size valved line for filling and a smaller size bypass including float valve and throttling valve for makeup.

z. To facilitate adjustments, the following should be provided:

   (1) Rate-of-flow indicators: In the discharge of all recirculating pumps, the makeup line and all jet subsystems.
   (2) Pressure gauges: In the discharge of all recirculating pumps, all jet subsystems, the suction of all recirculating pumps, and the inlets of all strainers. Check whether the pump suction and strainer inlet gauges should be compound type.

aa. Wind controls may be one- or two-stage, to reduce the fountain height (in one or two stages) or to shut the fountain down completely.

(1) To lower the height of the jets, the wind control either closes an automatically controlled throttling valve in the pump discharge a set amount or opens an automatically controlled throttling valve in the discharge suction bypass a set amount or both.
(2) To shut the fountain down completely, the wind control opens fully an automatically controlled throttling valve in the discharge suction bypass or shuts the pump off. In the latter case, a time-delay must be provided to prevent too-frequent operation of the pump.
(3) Automatically controlled throttling valves should be pneumatically operated through a PE switch.

bb. The filtration recirculation system should be provided with a hypochlorinator to prevent algae growth in the pool. For small pools and indoor pools, chlorination can be done by hand instead of mechanically.
cc. Supply piping should have a maximum velocity of 8 fps. Recirculating return piping to pumps should have a maximum velocity of 5 to 6 fps.
dd. Decorative fountain and pool vacuum cleaning is usually done with a portable electric or gasoline engine driven pump rather than a piped central system as with a swimming pool.

(1) Be sure to locate plaza drains of sufficient size in the general pool area to receive the discharge from the portable vacuum-cleaning pump.
(2) Remind the electrical project engineer to provide waterproof electric outlets in the general pool area to operate a portable vacuum-cleaning pump.
(3) The portable vacuum-cleaning pump, reel

of wire (if electric), and cleaning tools are usually supplied by the owner after occupancy; however, they can be supplied by the plumber if the owner desires. Where a portable electric vacuum-cleaning pump is provided, consideration must be given to the reel of extension wiring. Especially if the wiring is long, the reel becomes quite heavy and will need to be provided with wheels as is the pump. It is desirable to limit the wiring on the reel to a maximum of 125 ft to avoid a large reel size.

ee. When an aluminum pool is involved, all possible copper or copper alloys must be eliminated from the system.

(1) In addition, an easily replaceable sacrificial piece of aluminum piping at least 2 ft long, with at least one elbow (the more the better) and flanged or union ends should be provided in every line carrying water <u>to</u> the pool.
(2) Insulating bushings, couplings, unions, or flanges should be provided whenever any other metal contacts the aluminum.

2. Piping.

a. Fill and makeup water piping should be as specified for domestic water piping.
b. Since recirculated pool water is more corrosive than domestic water, it requires piping materials as good or better than that required for the domestic water.

(1) Inside the building, small piping should be copper tubing or red brass piping. Larger sizes should be bitumastic enamel (AWWA Specifications), epoxy, or plastic-lined steel pipe and malleable or cast-iron fittings.
(2) Underground piping may be polyvinyl

chloride (PVC) or polypropylene (PP) pipe and fittings.

(a) In small sizes, it may be copper tubing.
(b) In larger sizes, it may be cement-lined cast-iron or ductile-iron bell and spigot pressure pipe and fittings or asbestos cement (Transite) pressure pipe with cast-iron or ductile-iron fittings.

(3) Since most control valves are used for throttling, they should be ball valves in the smaller sizes and butterfly valves in the larger sizes.

c. Inside the building, plastic pipe and fittings may be used, if approved by your supervisor. Where piping is insulated, fire and smoke ratings of the insulation supercedes those of the pipe and fittings.

3. Equipment.

a. Filters.

(1) High-rate sand filters. Rate 15 gpm/ ft^2. Backwashed at recirculating rate. The type to use unless the size required is large.
(2) Vacuum diatomite filters. Rate 2 gpm/ ft^2. Require little water for backwash and minimum space. Due to fiberglass tank, they require no painting. The open tank can serve as the required air break and eliminate the need for a makeup tank.

(a) They require a body coat feeder.
(b) They require a pit to collect the used diatomite for manual removal. This is their major disadvantage.
(c) Where a vacuum-type filter is used, it is imperative that an accurate

section be drawn showing the elevations of the pool, the filter, and the filter room, since these vertical distances are critical in the selection of a filter and its control and must be made known to all manufacturers bidding on the job. In making these sections, consideration should be given to the foundation under the filter, because some filters require certain minimum height foundations due to their construction.

b. Pumps.

(1) The recirculating pumps should be flexible coupled horizontally split case, or close coupled vertically split case depending on the size. Where the local codes permit such an installation (not New York City) and the equipment room is rather remote from the pool, consideration should be given, as a matter of economy, to using a submersible pump (a deep well pump lying horizontal with an Anti-Vortex plate over it) in a depression in the bottom of the pool.

(2) The head on the pump in a bottom recirculating system with a pressure filter should be the sum of the following: The highest pressure required at the nozzle (or outlet) converted from psi to feet; the lift from the pool water level to the highest nozzle or trough behind a weir (if applicable); the discharge friction; the loss through the heater (if applicable); the loss through the filter (if applicable); the suction friction; and the loss through the strainer.

(3) The head on the pump in a system with a storage tank or vacuum filter should be the sum of the following: The highest pressure required at the nozzle (or outlet) converted from psi to feet; the lift from the pump to the pool water level, to the

trough behind a weir, or to the highest nozzle, whichever is applicable; the discharge friction; the loss through the filter (if applicable); the friction in the suction piping from the tank and the loss through the strainer (if applicable) converted from psi to feet.

(a) For vacuum diatomite filters, allow 20 ft for the loss through the filter.
(b) For high-rate sand filters, allow 15 to 20 psi converted from psi to feet for the loss through the filter.

(4) The initial selection of the jet-supply recirculating pump should be based on the required gpm and the head as calculated above. The final selection (and specification) should be based on the maximum that the required pump and motor can produce (to allow flexibility for field adjustment of fountain height); or if initial selection is the maximum that the pump and motor can produce, the next larger size pump and/or motor should be specified.
(5) NPSH requirements of recirculating pumps should be carefully checked.
(6) Normally fountain and pool recirculating pumps are manually operated, sometimes with additional start-stop buttons at a remote location.

(a) Occasionally the jet-supply recirculating pumps may be requested to be provided with time clock or wind control.
(b) As the filter recirculating pump should run 24 hr a day, provide start-stop control only at pump location.

(7) Pumps that are drawing from tanks should be provided with low water cutoff switches.

(8) Except for recirculating pumps pulling through diatomite filters, all recirculating pumps should be provided with full line size quick-opening basket strainers in their suction piping.

c. Heaters.

(1) Heaters should be shell and tube heat exchangers able to provide sufficient heat to the pool to prevent freezing in the winter.
(2) In small sizes, usually all the water is passed through the heater with the required temperature rise.
(3) At high flow rates (gpm), it is usually more economical to pass only part of the water through the heater at an increased temperature rise, then blend it back with the bypassed water.

d. Chlorination equipment.

(1) Hypochlorinators should be small, adjustable, positive displacement pumps (chemical feeders), built specially for the service, drawing sodium hypochlorite from a container and injecting it into the filter discharge piping to the pool after the last (hot-water-heater) connection.
(2) The hypochlorinator must be provided with a plastic container to hold the required sodium hypochlorite.

(a) Check that the units selected have a discharge pressure capability greater than that in the piping that they connect to.
(b) The table on page 9-43 gives suggested minimum-sized hypo chlorinators.

(3) Hypochlorinators should be interlocked with the filter recirculating pump so that they operate only when the pump is running.

System volume, gal	Suggested size, gph[a]
10,000	0.14
18,000	0.25
42,000	0.58
60,000	0.83
72,000	1.0
114,000	1.6
180,000	2.5
360,000	5.
720,000	10.
972,000	13.5
1,300,000	18.

[a]Capacity at maximum adjustment rating.

D. Compressed-air systems.

1. Principles of design.

a. There are four types of compressed-air systems, namely laboratory systems, hospital clinical (air used for patient treatment) systems, dental systems, and general building systems. Each must be a completely separate system.

(1) Laboratory systems.

(a) Provide outlets, connected to a central system, in laboratories as required by the program.
(b) The system should have rotary, liquid-ring-type compressors with receiver, filters, and constant-pressure valve, delivering relatively dry, clean, oil-free air at 55 psi, and with a maximum 5 psi pressure drop in the piping, 50 psi at the outlets.
(c) Provide lower-pressure requirements by pressure-reducing valves. Provide isolated higher-pressure requirements by isolated and/or subcentral compressor units of the required pressure.
(d) Where especially dry air is requested or piping runs through cold spaces (as in outside shafts), add an air dryer.

(e) Allow 1 cfm (free air) (scfm) for each laboratory outlet with simultaneous use factors as follows:

No. of outlets	Use factor, percent	Minimum cfm
1 - 2	100	
3 - 12	80	3
13 - 38	60	10
39 - 115	40	25
116 - 316	30	50
317 - 700	20	95
701 - 1,880	15	145
1,881 - 4,400	10	285
4,401 - 16,000	5	445
16,001 - 80,000	2	800

(f) The branches serving one or two classrooms should have 100 percent simultaneous use factor regardless of the number of outlets. The use factor for more than two classrooms may be 80 percent and thereafter twice the normal laboratory use factor if this is less than 80 percent.

(g) Simultaneous use factors must be used with judgement and modified to adapt to special conditions as they occur in the system.

(h) Demands with different simultaneous use factors should be carried separately in the calculations and added together only at the end.

(i) Piping should be sized to include future anticipated loads and to provide for normal flexibility in the laboratory areas.

(j) Minimum pipe sizes should be:

3/8-in branch to single or duplex outlet (short run)
1/2-in other branches
1/2-in riser
3/4-in main

(k) The outlets must be an appropriate

type for their locations and use, as specified.

(2) Hospital clinical systems.

(a) Provide outlets, connected to a central system, as required by the program, to serve patient treatment.
(b) Design these systems in accordance with the applicable requirements of medical gas systems including conformance with NFPA Standard 56F.
(c) The system should have rotary liquid ring-type compressors with receiver, dryers, filters and constant-pressure valve delivering dry, clean, oil-free air at 55 psi, and with a maximum 5-psi pressure drop in the piping, 50 psi at the outlets.
(d) Inlet piping to each compressor should be run independently through the roof as required by NFPA Standard 56F.
(e) Outlets are normally provided at least in the locations shown in the table on the next page and with the indicated required flows and simultaneous use factors. Check the program for the exact number and locations of all outlets.
(f) Carry the first outlet on the far end of a section of piping at full volume and apply simultaneous use factors to additional outlets only.
(g) Simultaneous-use factors must be disregarded when they produce a volume less than that of a single outlet.
(h) Simultaneous-use factors must be used with judgement and modified to adapt to special conditions as they occur in the system.
(i) Demands with different simultaneous-use factors should be carried

Location	Volume per outlet (cfm)	Simultaneous-use factor
Major Operating rooms (two outlets per operating room)	2	100
Minor Operating rooms	2	100
Emergency rooms[a]	2	100
Trauma rooms	2	100
Plaster room	1	50
Delivery rooms (two outlets per room)	2	100
Cystoscopy and Special Procedures	1	20
Recovery (one outlet per bed)	2	50
Intensive Care (ICU) rooms (one outlet per bed)	2	50
Coronary Care (CCU) rooms (one outlet per bed)	2	50
Patients rooms (bedside outlets)[b]	1	10
Nurseries (one per four bassinets)	1	20
Special Care Nurseries (one per incubator)	1	100
Examination and Treatment rooms	1	10
Surgical Preparation rooms	1	10
Blood Donor rooms	1	10
Anesthesia rooms	1	10
Cardiac and Heart Catheterization rooms	1	10
Inhalation Therapy	1	10
High-level Radioisotope rooms	1	10
Low-level Radiation rooms	1	10
X-ray rooms	1	10
Pharmacy	1	10
Sterile Supply	1	10
Autopsy	1	100

Number of outlets: one per room except as noted.

[a]All outlets in the emergency department (area) should have 100 percent simultaneous-use factor.

[b]Sometimes one outlet per bed and sometimes one outlet per two beds.

separately in the calculations and added together only at the end.

(j) Be generous in pipe sizing, particularly in the smaller sizes, because there are so many unknowns involved, and the use of compressed air in hospitals is on the increase.

(k) Arrange the risers so that patients' outlets on any floor of any wing are divided between at least two risers.

(l) A major operating suite should be served by more than one riser.

(m) Provide the branches feeding outlets grouped with other medical gas outlets with ball valves mounted in a common multiple-valve box.

(n) The minimum pipe sizes should be:

3/8-in drops feeding a single outlet
1/2-in other branches
3/4-in riser
3/4-in main

(o) For outlets occurring in conjunction with medical gas outlets, see Section H.
(p) Both central and local alarms should be provided as required for medical gases.

(3) Dental systems.

(a) Provide outlets in dental-chair pedestals connected to a central system, as required by the program for operating dental drills.
(b) Design these systems in accordance with the applicable requirements of medical gas systems (Section H), including conformance with NFPA Standard 56F.
(c) The system should have rotary liquid ring or other oilless-type compressors with receiver, dryer, filters, and constant-pressure valve; delivering clean, dry, oil-free air at 85 to 100 psi, with a maximum 5 psi pressure drop in the piping.
(d) Contact the architect, owner, and/or tool manufacturer about the type of tools being considered and their volume and pressure requirements.
(e) The minimum pipe sizes should be:

3/8-in branch to a single outlet (short run)
1/2-in other branches
3/4-in riser
3/4-in main

(4) General building systems.

(a) All other compressed-air requirements.

(b) Provide compressed air of required quality in the required amounts and at the required pressures to satisfy the building's needs.

(c) Check with the architect, owner, and/or equipment manufacturer about volumes, simultaneous-use factors, and pressures required.

2. General.

a. Data in this section is based on 14.7-psia atmospheric pressure. In locations significantly different, cfm's, friction losses, and equipment selections must be adjusted accordingly.
b. Make ample provision in the central plant and in the piping for anticipated future building expansion.
c. Locate central compressed-air plants below the areas they serve with up-feed risers.
d. Compressed-air piping should pitch back to the central plant wherever possible. Provide drains at the low points and trapped sections of the mains. The drains should be manual petcocks or automatic drain traps depending on the extent of the system served.
e. All risers and branches should be valved. Provide mains with sectionalizing valves at strategic locations. All connections for future extension should be valved.

3. Pipe sizing.

a. Size piping on the basis of the cfm loads and simultaneous-use factors given above.
b. It should be noted that all cfm referred to in this section are free air at atmospheric pressure and are not the actual volumes of the compressed air in the piping which will be less due to compression and will vary depending on the pressure in the piping.
c. Base pipe-size selection in each case on the more stringent of the following requirements:

(1) Maximum friction loss rate of 1 psi per 100 ft

(2) Maximum friction loss to the farthest outlet of 5 psi

(3) Maximum velocity of 4,000 ft/min

d. Table 9-1 gives the friction losses in 50 psi compressed-air piping. The actual run of the piping must be converted to equivalent developed length by adding fitting and valve allowances before using the tables. The maximum

Table 9-1 Pressure Loss (psi) per 100 ft in 50 psi Compressed-Air Piping

	Normal pipe size, in								
cfm of free air	1/2	3/4	1	1-1/4	1-1/2	2	2-1/2	3	4
5	0.30	0.03	0.01						
10	1.15	0.18	0.05	0.01					
15		0.40	0.11	0.03					
20		0.69	0.20	0.05	0.02				
25		1.14	0.31	0.07	0.03				
30			0.44	0.10	0.05				
35			0.61	0.14	0.06				
40			0.80	0.18	0.08				
45			1.00	0.23	0.10	0.03			
50				0.29	0.13	0.04			
60				0.42	0.18	0.05			
70				0.56	0.25	0.07	0.03		
80				0.74	0.33	0.09	0.03		
90				0.93	0.41	0.11	0.04		
100				1.15	0.51	0.14	0.05		
110					0.62	0.17	0.06		
120					0.73	0.20	0.08		
130					0.86	0.23	0.09	0.03	
140					1.00	0.27	0.11	0.03	
150						0.31	0.12	0.04	
175						0.42	0.16	0.05	
200						0.54	0.21	0.07	
225						0.69	0.29	0.08	
250						0.85	0.33	0.10	
275						1.03	0.40	0.13	
300							0.48	0.15	
325							0.56	0.18	
350							0.65	0.20	
375							0.74	0.23	
400							0.84	0.27	
450							1.06	0.32	
500								0.42	
550								0.50	0.12
600								0.60	0.14
650								0.70	0.17
700								0.82	0.19
750								0.94	0.22
800								1.06	0.25
850									0.28
900									0.32
950									0.36
1,000									0.39
1,100									0.48
1,200									0.57
1,300									0.67
1,400									0.77
1,500									0.89
1,600									1.00

capacity not to exceed 4,000 fpm: 4 in - 1,600 cfm.

e. For friction losses in other than 50 psi systems, use Table 9-2 or consult standard friction tables for the pressures involved.

f. See Calculation Form N, Section 2, for sizing piping.

4. Piping.

a. Piping carrying air for patient treatment, regardless of size, should be type L copper tubing with wrought copper or cast brass fittings with silver brazed joints, or red brass pipe with threaded cast brass fittings. All pipe, fittings, and valves should be specially washed for medical gas use and protected from contamination thereafter.

b. Except as otherwise noted, piping 2 in and smaller should be type L copper tubing with cast brass or wrought copper solder joint fittings, or red brass pipe with threaded cast brass fittings.

c. Except as otherwise noted, piping 2-1/2 in and larger can be galvanized steel pipe with galvanized threaded malleable iron fittings.

d. For general building systems, small piping may be galvanized steel. In some industrial systems, piping may be welded steel.

e. Buried piping must be adequately protected against frost, corrosion, and physical damage by installation within a pipe or conduit, with proper cover, and adequate corrosion protective coating.

f. Control (shutoff) valves should be ball valves.

g. Check valves should normally be spring-loaded type.

5. Equipment.

a. Size air compressors to carry the peak anticipated load with one compressor out of operation.

(1) Where future loads are involved, if the

Table 9-2 Air-Pressure Loss, psi in 100 ft of Clean Commercial Steel Pipe

CFM of Free Air	Nominal Pipe Diameter														
	½ Inch			¾ Inch			1 Inch			1 ¼ Inches			1 ½ Inches		
	80 lb.	100 lb.	125 lb.	80 lb.	100 lb.	125 lb.	80 lb.	100 lb.	125 lb.	80 lb.	100 lb.	125 lb.	80 lb.	100 lb.	125 lb
10	.46	.38	.31	.11	.09	.08	.04	.03	.02	.0086	.0071	.0058			
20	1.74	1.42	1.17	.41	.34	.28	.13	.10	.08	.032	.026	.021	.014	.012	.010
30	3.84	3.13	2.54	.90	.74	.60	.28	.23	.19	.068	.056	.046	.031	.026	.021
40	6.93	5.55	4.53	1.55	1.28	1.05	.46	.38	.31	.116	.096	.079	.053	.044	.036
50	10.7	8.65	7.01	2.42	2.00	1.62	.73	.60	.49	.18	.146	.120	.081	.067	.055
60				3.47	2.84	2.33	1.02	.84	.69	.25	.21	.17	12	.095	.078
70				4.73	3.85	3.14	1.36	1.12	.92	.34	.28	.23	.16	.13	.10
80				6.14	5.01	4.08	1.76	1.44	1.18	.44	.36	.30	.20	.16	.14
90				7.75	6.40	5.17	2.23	1.85	1.49	.55	.45	.37	.25	.20	.17
100				9.62	7.80	6.33	2.69	2.21	1.81	.66	.55	.45	.30	.25	.20
125				15.5	12.4	9.8	4.18	3.41	2.79	1.03	.85	.69	.46	.38	.32
150				23.0	18.1	14.4	5.75	4.91	3.99	1.47	1.20	.99	.65	.54	.44
175							8.10	6.80	5.45	2.00	1.64	1.32	.90	.73	.60
200							10.9	8.79	7.11	2.58	2.12	1.73	1.15	.95	.78
250										4.05	3.30	2.67	1.82	1.48	1.20
300										5.78	4.71	3.83	2.55	2.10	1.72
350										7.90	6.45	5.15	3.53	2.86	2.35
400										10.3	8.30	6.74	4.53	3.70	3.03
450													5.80	4.65	3.80
500													7.12	5.79	4.71

CFM of Free Air	Nominal Pipe Diameter														
	2 Inches			2 ½ Inches			3 Inches			4 Inches			5 Inches		
	80 lb.	100 lb.	125 lb.	80 lb.	100 lb.	125 lb.	80 lb.	100 lb.	125 lb.	80 lb.	100 lb.	125 lb.	80 lb.	100 lb.	125 lb.
50	.024	.020	.016	.010	.008	.007									
60	.033	.027	.022	.014	.011	.009									
70	.044	.036	.030	.018	.015	.012									
80	.056	.046	.038	.023	.019	.015									
90	.070	.058	.048	.028	.024	.020									
100	.084	.069	.057	.035	.029	.023	.012	.010	.008						
125	.130	.107	.088	.055	.043	.036	.018	.015	.012						
150	.19	.15	.12	.074	.061	.050	.025	.021	.017						
175	.25	.20	.17	.099	.081	.067	.034	.028	.022						
200	.31	.26	.21	.128	.105	.086	.043	.036	.029						
250	.49	.40	.33	.195	.160	.131	.065	.054	.044	.017	.014	.011			
300	.70	.57	.47	.27	.23	.19	.092	.075	.062	.024	.020	.016			
350	.94	.77	.64	.37	.31	.25	.124	.101	.083	.032	.026	.022			
400	1.20	.99	.81	.48	.40	.33	.160	.131	.108	.041	.034	.028			
450	1.55	1.27	1.05	.60	.50	.41	.20	.165	.135	.051	.042	.035			
500	1.91	1.56	1.29	.75	.62	.50	.25	.20	.17	.062	.051	.042			
600	2.75	2.23	1.83	1.08	.89	.73	.35	.29	.24	.089	.073	.060			
700	3.67	3.00	2.45	1.43	1.18	.97	.48	.39	.32	.119	.098	.081			
800	4.90	4.00	3.23	1.87	1.54	1.25	.61	.50	.41	.154	.126	.104			
900	6.20	5.05	4.10	2.35	1.95	1.57	.77	.63	.54	.193	.159	.131			
1000	7.62	6.20	5.04	2.89	2.37	1.94	.94	.78	.64	.23	.19	.16	.075	.062	.051
1200	11.4	9.05	7.45	4.21	3.45	2.78	1.37	1.12	.92	.34	.28	.23	.105	.088	.072
1500	18.3	14.5	11.7	6.62	5.39	4.38	2.15	1.73	1.43	.52	.43	.35	.162	.135	.110
2000				12.0	9.66	7.80	3.77	3.09	2.52	.91	.75	.61	.28	.24	.192
2500							6.00	4.85	4.00	1.43	1.16	.95	.44	.37	.30
3000							8.60	6.98	5.67	2.05	1.65	1.35	.63	.52	.43
3500							12.0	9.65	7.80	2.78	2.25	1.84	.85	.70	.58
4000							15.9	12.6	10.2	3.55	2.91	2.37	1.11	.91	.75
4500										4.60	3.70	3.00	1.40	1.16	.95
5000										5.68	4.55	3.70	1.73	1.43	1.17
6000										8.12	6.58	5.34	2.45	2.01	1.65
7000													3.40	2.79	2.28
8000													4.46	3.63	2.89
9000													5.65	4.60	3.76
10,000													6.95	5.65	4.59

CFM of Free Air	Nominal Pipe Diameter		
	6 Inches		
	80 lb.	100 lb.	125 lb.
1500	.063	.052	.043
2000	.110	.091	.075
2500	.170	.140	.115
3000	.24	.20	.164
3500	.33	.27	.22
4000	.43	.35	.29
4500	.54	.44	.36
5000	.67	.55	.45
6000	.94	.77	.63
7000	1.30	1.06	.87
8000	1.70	1.39	1.13
9000	2.15	1.76	1.43
10000	2.61	2.14	1.76
11000	3.23	2.63	2.14
12000	3.87	3.13	2.55

Lengths Other Than 100 Feet

The friction loss in pipe lengths shorter than 100 feet may be calculated proportional to the length. That is: for 50 feet, ½ the tabular figure; for 25 feet, ¼ the tabular figure, etc.

In pipe runs of more than 100 feet, the proportional method may be used providing the resultant friction loss does not exceed 9 or 10 psi. If it is greater, a more accurate check may prevent undersizing of pipes. Use the following method, based on the Fanning Equations from which these tables were derived:

Calculate first a conversion figure "K":

$$K = \frac{F_{100}(P - \tfrac{1}{2} F_{100})}{100}$$

Then: $F_L = P - \sqrt{P^2 - 2KL}$

Where: F_{100} = tabular friction loss figure for 100 ft.
F_L = friction loss for length "L"
L = length of pipe in feet
and P = upstream pressure in pipe, psia

Source: Reprinted with permission of Ingersoll Rand Co.

proportions are workable, a three compressor installation may be the more practical with the third compressor installed later.

(a) This will require a smaller standby unit.
(b) Also a three compressor installation will provide a smaller unit for operation during low-demand periods.

b. Compressors for laboratory, hospital clinical and dental systems should be rotary liquid ring type delivering nonpulsating clean, oil-free, moisture-free air.

(1) If pressures required are over 100 psi, oilless piston type will have to be used.
(2) In large installations, where chilled water is always available, recirculated seal water systems can be considered to conserve water usage.
(3) Use the following minimum sizes for the air compressor intake from the roof:

cfm free-air capacity	Size, in
50	2-1/2
70	3
110	3
210	4
400	5

(4) Because inlet pressure (friction loss in the inlet piping), if over 1/2 psi, affects the manufacturer's published air-compressor rating tables, inlet-pressure conditions, where the inlet piping has to be run up through the roof, should be checked with the manufacturer before finalizing equipment selections.
(5) Since these air compressors have minimum seal water pressure requirements, check this required pressure and connect

to a building system of at least that pressure.

c. For general building systems, depending on the use of the air and the pressures required, compressors may be rotary liquid ring type, oilless piston type, or oil-lubricated piston type, either air-cooled or water-cooled depending on the size. They are usually provided with after coolers.

d. Air dryers: hospital clinical and dental systems, and elsewhere when absolutely dry air is required, should be provided with air dryers.

(1) Normally these should be refrigerated dryers giving a pressure dew point of 35°F.

(2) Where a lower dew point is required, provide a combination refrigerated and desiccant dryer.

(3) Where dryers are provided, they should be complete with filters, and the separate pipeline filters omitted.

(4) Consider small local units for isolated requirements.

(5) In small sizes, compressor, dryer, and receiver are available as a package unit.

e. The size of the receiver (tank), inlet mufflers, inlet filters, and constant-pressure valve should be as recommended by the compressor manufacturer for the ultimate installation.

f. Filters on the discharge main from the central plant should be duplex absorption type, each sized for approximately one half to two thirds of the ultimate load. If proportions are workable, it may be desirable to use three filters with one future. Provide automatic drain traps on the filters.

E. Vacuum air systems.

1. Principles of design.

a. Provide inlets connected to a central system in laboratories as required by the program.

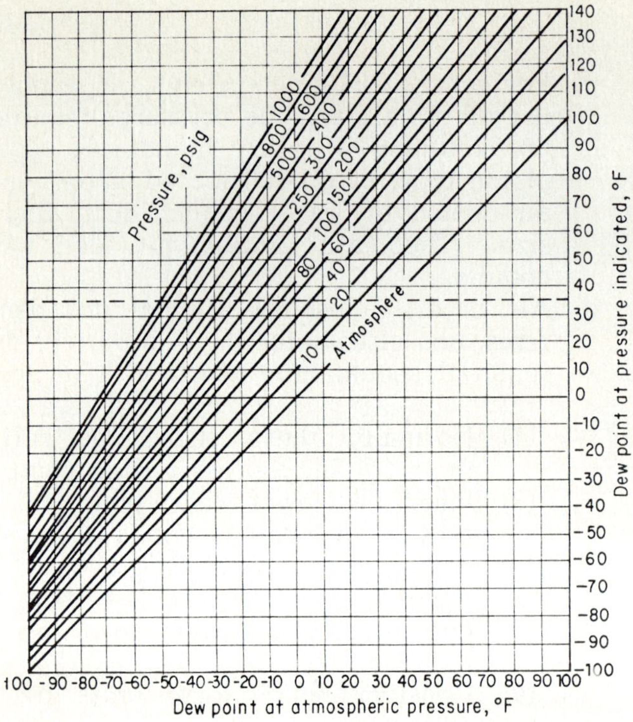

Dew point conversion:

To obtain the dew point temperature expected if the gas were expanded to a lower pressure proceed as follows:

1. Using "dew point at pressure", locate this temperature on scale at right hand side of chart
2. Read horizontally to intersection of curve corresponding to the operating pressure at which the gas was dried
3. From that point read vertically downward to curve corresponding to the expanded lower pressure
4. From that point read horizontally to scale on right hand side of chart to obtain dew point temperature at the expanded lower pressure
5. If dew point temperatures at atmospheric pressure are desired, after step 2 above read vertically downward to scale at bottom of chart which gives "Dew Point at Atmospheric Pressure"

Fig. 9-2 Dew point conversion chart. Hankison Corporation.

b. Provide inlets connected to a central system as required by the program in hospitals to serve patient treatment. Connect all inlets that serve patient treatment to a completely separate system from those serving the laboratory inlets.

c. Make ample provision in the central plants and in the piping for anticipated future building expansion.

d. Patient vacuum air inlets are normally provided at least in the locations shown on pages 9–55

Location	cfm per inlet (at 15-in Hg)	Simultaneous-use factor
Operating room open heart, organ transplant, etc. (two per room)	3.5	100
Major Operating room (two per room)	2	100
Minor Operating room (two per room)	2	100
Cystoscopy and Special Procedures (two per room)	1	40
Emergency[a] (two per room)	2	100
Trauma room	1	100
Plaster (Fracture room)	1	100
Delivery room (two per room)	2	100
Recovery first inlet per bed	3	50
Recovery second inlet per bed	1	50
Recovery additional inlets per bed	1	10
ICU and CCU first inlet per bed	3	50
ICU and CCU second inlet per bed	1	50
ICU and CCU additional inlets per bed	1	10
Patient rooms surgical[b]	1	50
Sometimes one inlet per bed		
Sometimes one inlet per two beds		
Patient rooms medical[b]	1	10
Sometimes one inlet per bed		
Sometimes one inlet per two beds		
Labor rooms	1	20
Nurseries (one per four bassinets)	1	10
Special Care Nurseries (one per incubator)	1	40
Examination and Treatment rooms	1	10
Operating room bed holding areas	1	10
Surgical Preparation rooms	1	10
Blood Donor rooms	1	10
Anesthesia Work rooms	1	10
Cardiac and Heart Catheterization rooms	1	10
Deep Therapy rooms	1	10
Inhalation Therapy rooms	1	10
Electroencephalogram (EEG) rooms	1	10
Electrocardiogram (ECG) rooms	1	10
Electromyogram (EMG) rooms	1	10
Fluoroscopy rooms	1	10
High-level Radioisotope rooms	1	10
Low-level Radiation rooms	1	10
X-ray rooms	1	10
Bronchography room	1	100
Autopsy room	2	100
Central Supply room	1	10
Pharmacy	1	40

Number of inlets: one per room except as noted.

[a]All inlets in the emergency department (area) should have 100 percent simultaneous-use factor.

[b]Where patient rooms are interchangeable (surgical or medical), use 50 percent simultaneous-use factor for the first four rooms on the far end of the section of piping and 20 percent thereafter.

and 9-56 and with the indicated required flows and simultaneous-use factors. Check the program for the exact number and locations of all inlets.

e. Allow 1 cfm at 15 in Hg for each laboratory inlet with simultaneous use factors as follows:

No. of inlets	Use factor	Minimum cfm
1 to 4	100	
6 to 12	80	5
13 to 33	60	10
34 to 80	50	21
81 to 150	40	40
151 to 315	35	61
316 to 565	30	111
566 to 1,000	25	171
1,001 to 2,175	20	251
2,176 to 4,670	15	436
4,671 up	10	701

The branches serving one or two classrooms must have 100 percent simultaneous-use factor regardless of the number of outlets. Use factor for more than two classrooms may be 80 percent, and thereafter twice the normal laboratory-use factor if this is less than 80 percent.

f. Simultaneous-use factors must be used with judgement and modified to adapt to special conditions as they occur in the system.

g. Demands with different use factors should be carried separately in the calculations and only added together at the end.

h. The central vacuum air plant should be located below the areas served with upfeed risers.

i. The vacuum pump discharge should be run up through the roof remote from all compressed air and other intakes, windows, etc.

j. The vacuum air piping should pitch back to the central plant wherever possible. Valved drain pockets and drain cocks should be provided at low points and trapped sections of the main.

k. Plugged cleanout connections should be strategically located throughout the system to allow a means for removing stoppages. These can be created by using a tee instead of an elbow or a cross instead of a tee.

l. The branches to risers should connect to the riser with a drop leg to facilitate the draining of the contents of the piping back to the central equipment.

m. The branch piping to inlets should feed up to the inlets rather than down to the inlet wherever possible. Wherever downfeed branches are unavoidable, they should be connected to the top of the main branch to minimize the draining back to the inlet of the contents of the piping in the event of even momentary failure of the vacuum in the system.

n. All risers and branches should be valved. Mains should be provided with sectionalizing valves at strategic locations. All connections for future extension should be valved.

o. The branches feeding inlets grouped with other medical gas outlets should be provided with shutoff valves mounted in a common multiple-valve box. For branches 1-1/4 in and larger, valves may have to be in separate adjacent boxes.

p. The risers should be so arranged that patients' inlets on any floor of any wing are divided between at least two risers.

q. A major operating suite should be served by more than one riser.

r. The laboratory inlets should be the appropriate type for their location and use, as specified.

s. For inlets occurring in conjunction with medical gas outlets, see Section H.

t. Both central and local alarms should be provided for hospital clinical systems as required for medical gases.

u. Data in this section is based on 14.7 psia atmospheric pressure. In locations significantly different, cfm's, friction losses, and equipment selections must be adjusted accordingly.

2. Pipe sizing.

a. Size piping on the basis of the cfm loads and simultaneous-use factors given above.

b. Note that the actual volume (cfm) of the air in the piping at any point is greater than the volume (cfm) of the room air (at atmospheric pressure) entering the system, due to expansion under vacuum (subatmospheric) conditions.

   (1) For 15 in Hg, the ratio is 2 to 1.
   (2) For 19 in Hg, the ratio is 2.7 to 1.
   (3) Be sure that the inlet requirments, the friction-loss table, and the vacuum pump rating table all use the same cfm basis.
   (4) The data given herein is all compatible; all cfm at 15 to 19 in Hg.

c. Size piping to include future anticipated loads and to provide for normal flexibility in laboratory areas.
d. Base pipe-size selection in each case on the more stringent of the following requirements:

   (1) Maximum friction loss rate of 1 in Hg per 100 ft
   (2) Maximum friction loss to the farthest outlet of 4 in Hg
   (3) Maximum velocity of 5,000 ft

e. Table 9-3 gives the friction losses in the vacuum air piping: Convert the actual run of the piping to equivalent developed length by adding fitting and valve allowances before using the table.
f. The following is the maximum capacity not to exceed 5,000 fpm:

2 in	110 cfm
2-1/2 in	170 cfm
3 in	250 cfm
4 in	435 cfm
5 in	700 cfm
6 in	1,000 cfm

g. In large systems, consideration may be given to the fact that friction losses in piping 2 in and larger are less than average; therefore, some of this saving in friction can be used up in smaller

Table 9-3 Pressure Loss per 100 ft of Pipe, in Hg

cfm[a]	Nominal pipe size, in 3/4	1	1-1/4	1-1/2	2	2-1/2	3	4	5	6
1	0.02									
2	0.08	0.02								
3	0.16	0.04								
4	0.24	0.06	0.02							
5	0.36	0.08	0.02							
6	0.48	0.12	0.04							
7	0.60	0.16	0.04							
8	0.76	0.20	0.06							
9	0.92	0.24	0.08							
10	1.10	0.28	0.10	0.04						
15		0.60	0.18	0.08	0.02					
20		1.04	0.30	0.14	0.04					
25		1.12	0.47	0.22	0.06	0.02				
30			0.68	0.30	0.08	0.02				
35			0.90	0.40	0.10	0.04				
40			1.10	0.50	0.14	0.04	0.02			
45				0.62	0.16	0.06	0.02			
50				0.74	0.20	0.06	0.03			
60				1.00	0.26	0.10	0.04			
70				1.32	0.36	0.12	0.05	0.02		
80					0.50	0.16	0.06	0.02		
90					0.62	0.20	0.08	0.02		
100					0.76	0.24	0.10	0.03		
125					1.10	0.34	0.15	0.04		
150						0.55	0.23	0.06		
175						0.70	0.30	0.08		
200							0.39	0.08		
250							0.58	0.13		
300								0.20		
350								0.28		
400								0.35	0.10	
450								0.45	0.13	
500									0.17	
600									0.24	
700									0.30	0.10
800										0.12
900										0.16
1,000										0.20
1,100										0.24

[a]cfm is expanded cfm at 15 to 19 in Hg. NOTE: To use the NCG or Puritan-Bennett calculators, the cfm used must be cfm at atmospheric (room) pressure (scfm) (free air).

sizes, still keeping the total under 4 in Hg. However, the loss in any pipe must be kept at 1 in Hg/100 ft or less.

h. Minimum pipe sizes should be:

1/2-in branch to a single or duplex outlet (short run)
3/4-in other branches
3/4-in riser
1-in main

i. See Calculation Form O, Section 2, for sizing piping.

3. Piping.

   a. Piping 2 in and smaller should be type L copper tubing with wrought copper long turn solder joint fittings or red brass pipe with threaded cast brass long turn fittings.
   b. Piping 2-1/2 in and larger should be galvanized steel pipe with threaded galvanized cast-iron long turn drainage pattern fittings.
   c. Piping should be designed using Y or TY fittings instead of tees.
   d. Control (shut off) valves should be ball valves.
   e. Check valves should be soft-seated swing type.

4. Equipment.

   a. The vacuum pumps should be rotary liquid ring type.
   b. The vacuum pumps should be sized to carry the peak anticipated load at 20 in Hg with one pump out of operation. Where future loads are involved, if the proportions are workable, a three pump installation may be more practical with the third pump installed later.

      (1) This will require a smaller standby unit.
      (2) Also a three pump installation will provide a smaller unit for operation during low demand periods.

   c. Use the following minimum sizes for the vacuum pump discharge piping to the roof:

cfm capacity	Size, in
12	1-1/4
23	1-1/2
40	2
72	2-1/2
130	3
160	4
190	4
350	5
525	5

d. Because back pressure (friction loss in the discharge piping), if significant, affects the manufacturer's published vacuum pump rating tables, back-pressure conditions on each project should be checked with the manufacturer before finalizing equipment selections.
e. Vacuum air at the central plant should be maintained at 20 in Hg.
f. The size of the vacuum tank should be as recommended by the pump manufacturer for the ultimate installation.
g. In large installations, where chilled water is always available, recirculated seal water systems can be considered to conserve water usage.
h. In systems serving patient treatment and in medical research laboratory systems, filters may be required in the discharge. Filters should be in duplex. The filter pressure drop must be added to pipe friction losses in calculating discharge losses.

F. Oral vacuum systems.

1. Principles of design.

a. Connect oral vacuum inlets in dental-chair pedestals to a central oral vacuum system.
b. Make ample provision in the equipment and piping for anticipated future building expansion.
c. Calculate the system on the basis of 10 cfm inlet at 5-1/2 in Hg with normally a 100 percent use factor. A lower total system use factor should only be used if information from the owner on intended use justifies it. Vacuum producers and separator pumps should be duplex each full size.
d. The central vacuum plant should be centrally located in the basement with upfeed risers.
e. The piping should pitch back to the separator.
f. The branch piping to inlets should feed up to the inlet rather than down to the inlet, wherever possible. Wherever downfeed branches are unavoidable, they should be connected to the top of the main or main branch to minimize the draining back to the outlet of the contents of

the piping in the event of even momentary failure of the vacuum in the system.

g. All risers and branches should be valved. All connections for future extension should be valved.

h. The vacuum producer discharge should be connected to the chimney or flue, or run up through the roof.

i. Plugged cleanout connections should be strategically located throughout the system to allow a means for removing stoppages. These can be created by using a TY instead of an elbow.

j. Data in this section is based on 14.7 psia atmospheric pressure. In locations significantly different, cfm's, friction losses, and equipment selections must be adjusted accordingly.

2. Pipe sizing.

a. Size piping on the basis of the cfm loads and normally a 100 percent use factor. Minimum pipe size should be 2 in.

b. Note that the actual volume (cfm) of the air in the piping at any point is greater than the volume (cfm) of the room (at atmospheric pressure) due to expansion under air vacuum (subatmospheric) conditions entering the system.

(1) Approximate expansion ratios:

in Hg	Ratio
4	1.15
5	1.2
6	1.25
7	1.3
8	1.35

(2) Be sure that the inlet requirements, the friction-loss table, and the vacuum producer rating table use the same cfm basis.

(3) The data given herein is all compatible, all cfm (known as scfm) at atmospheric pressure.

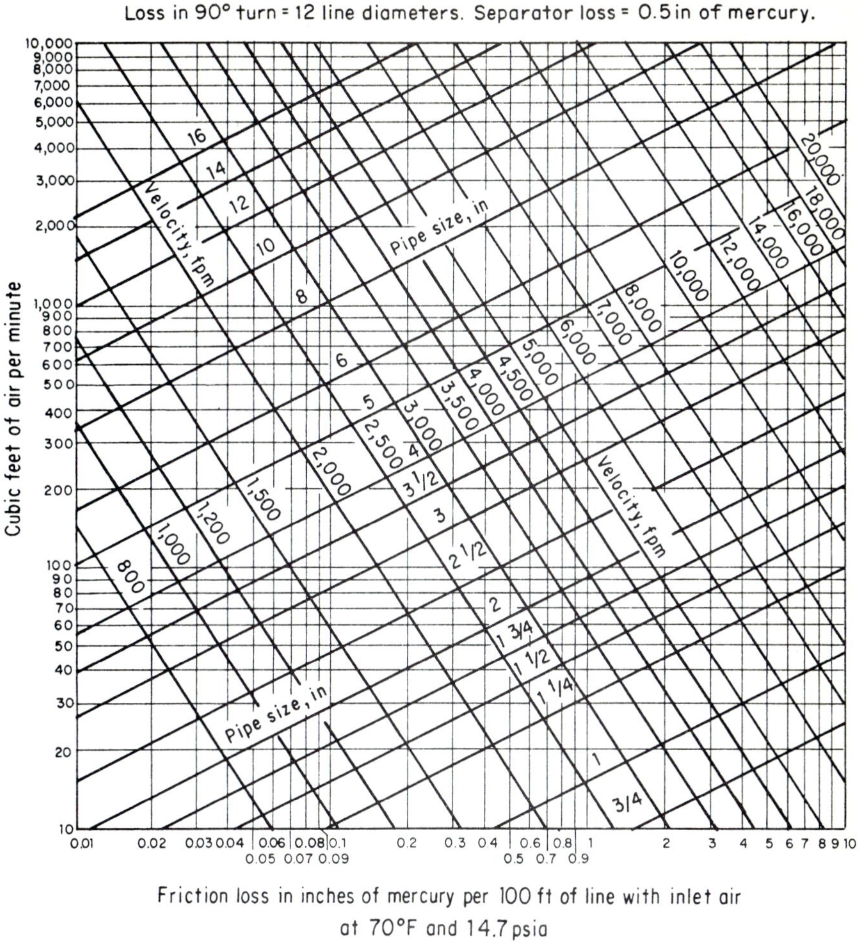

Fig. 9-3 Friction losses in oral vacuum piping. Compliments of the Spencer Turbine Co., Windsor, Conn.

c. Figure 9-3 gives the friction losses in the oral vacuum piping.

   (1) Convert the actual run of piping to equivalent developed length before using the table.
   (2) Long mains may have to be oversized to cut down excessive friction losses.
   (3) The friction in the discharge piping should be kept minimal.

3. Piping.

   a. 2-in piping should be type L copper tubing with wrought copper long turn solder joint fittings or red brass pipe with threaded cast brass long turn fittings.
   b. Piping 2-1/2 in and larger should be galvanized steel pipe with threaded galvanized long turn cast iron drainage fittings.
   c. Piping should be designed using Y or TY fittings insteads of tees.
   d. Control valves should be wafer or top-entry-type ball valves.

4. Equipment.

   a. Vacuum producers.

      (1) The vacuum producers should be duplex multistage centrifugal type with automatic control.
      (2) The vacuum head required for the system should be the sum of the following: inlet loss (5-1/2 in Hg), the system piping losses, the separator loss (0.5 in Hg) and the discharge loss.
      (3) Provide rubber flexible connections in both the suction and discharge piping at the vacuum producers.
      (4) Provide a check valve and blast gate in each discharge of duplex units, and a control valve in the suction.
      (5) Provide vacuum producers with surge control devices with silencer or use vacuum producers specially designed to overcome surges.
      (6) Where noise may be a factor, provide silencers on the discharge.

   b. Separator.

      (1) The separator should be a centrifugal hospital type with duplex drain pumps.

         (a) Pump capacity: 0.3 gpm per chair in use.

(b) Pump control, automatic.
(c) Provide a 6-in high curb around the separator with a floor drain capable of handling the pump-discharge rate.

G. Vacuum cleaning systems.

1. Principles of design.

a. Provide vacuum cleaning systems only when requested by the owner and/or architect. Their use is rare at the present time.
b. Vacuum cleaning systems are not covered by code.
c. Design the systems with the risers so located that all of the areas to be cleaned can be reached using not over 50 ft of 1-1/2-in hose. In hospitals, to minimize dangerous obstruction to corridor traffic (hose laying on the floor across the corridor), provide inlets on both sides of the corridor.
d. Assume an operator can clean 3,000 ft^2/h.
e. Make ample provision in the equipment and the piping for anticipated future building expansion.
f. In hospitals, use silent-type inlet valves in patient areas.
g. The hose and cleaning tools are usually provided by the plumber.
h. The central vacuum cleaning plant should be centrally located in the basement with upfeed risers.
i. The piping should pitch back to the separator.
j. The branch piping to inlets should feed up to the inlet rather than down to the inlet wherever possible.
k. The vacuum producer discharge should be connected to the chimney or flue, or run up through the roof.
l. Plugged cleanout connections should be strategically located throughout the system to allow a means for removing stoppages. These can be created by using a TY instead of an elbow.
m. Data in this section is based on 14.7 psia atmospheric pressure. In locations significantly different, cfm's, friction losses, and equipment selections must be adjusted accordingly.

2. Pipe sizing.

a. Size piping on the number of simultaneous operators expected to use the system, not the number of inlets. Minimum pipe size should be 2 in.

b. Each operating inlet uses 70 cfm. Use the following table for pipe sizing.

Size, in	Maximum number of operators
2	1
2-1/2	2
3	3
3-1/2	4
4	5
5	8
6	12
8	20

c. It should be noted that the actual volume (cfm) of air in the piping at any point is greater than the volume (cfm) of the room air (at atmospheric pressure) entering the system, due to expansion under vacuum (subatmospheric) conditions.

(1) Approximate expansion ratios:

(1) Approximate expansion ratios

in Hg	Ratio
2	1.1
3	1.1
4	1.15
5	1.2
6	1.25
7	1.3
8	1.35
9	1.4

(2) Be sure that the inlet requirements, the

friction loss table, and the vacuum producer rating table use the same cfm basis.

(3) The data given herein is all compatible, all cfm (known as scfm) at atmospheric pressure.

d. Figure 9-4 gives the friction losses in the vacuum cleaning piping.

(1) Convert the actual run of piping to

Hose loss = 1-1/2 times line loss. Loss in 90° turn = 12 line diameters. Separator loss = 0.5 in of mercury. Suction at tool = 2.0 in of mercury (average conditions)

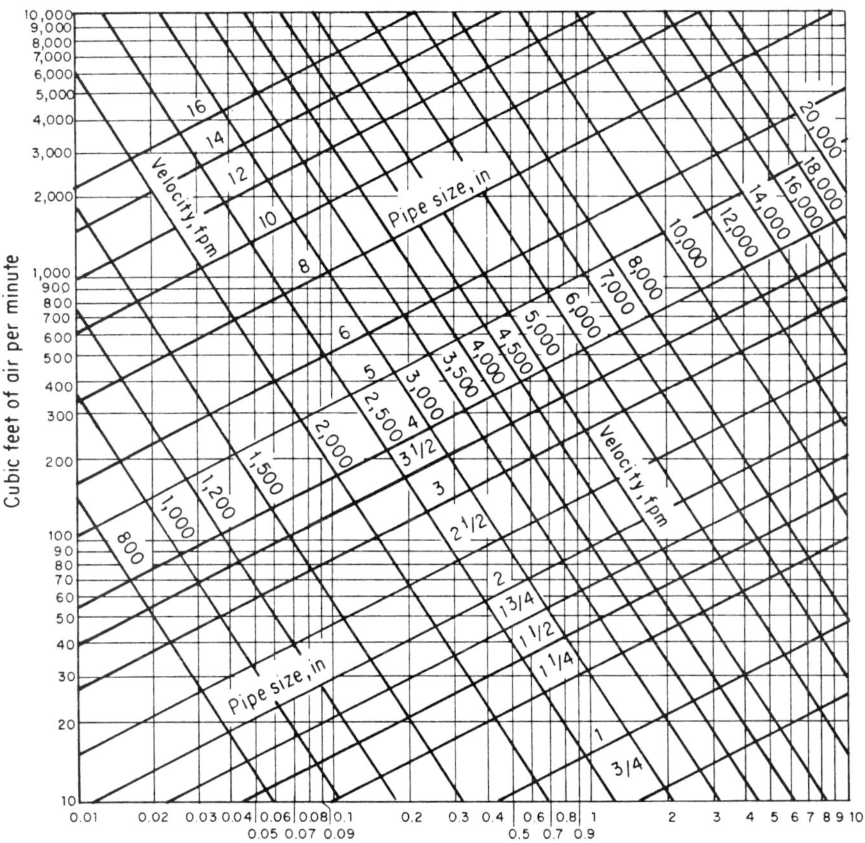

Friction loss in inches of mercury per 100 ft of line with inlet air at 70°F and 14.7 psia

Fig. 9-4 Friction losses in vacuum cleaning piping. Compliments of the Spencer Turbine Co., Windsor, Conn.

equivalent developed length before using the above table.

(2) Long mains may have to be oversized to cut down excessive friction losses.

(3) The friction loss in the discharge piping should be kept minimal.

3. Piping.

a. Normally piping should be black steel tubing and long turn fittings with brazed joints.
b. In New York City, due to union rules, piping is standard-weight steel pipe with threaded long turn cast-iron drainage fittings.
c. Design piping using Y or TY fittings instead of tees.

4. Equipment.

a. Vacuum producers.

(1) The vacuum producer should be multistage centrifugal type. Control should be manual.

(2) The vacuum head required for the system should be the sum of the following: the inlet loss (2 in Hg), the hose loss (1-1/2 times the loss in 1-1/2-in pipe), the system piping losses, the separator loss (0.5 in Hg for each separator), and the discharge loss.

(3) Provide rubber flexible connections in both the suction and discharge piping of the vacuum producer.

(4) Provide a check valve and blast gate in each discharge of duplex units.

(5) Where required (check with manufacturer), provide surge control device with silencer.

(6) Where noise may be a factor, provide silencer on the dicharge.

b. Separators.

(1) Normally for systems of six operators or less, only a tubular bag-type separator is

provided. For systems of eight operators or more, both a centrifugal-type separator (first in line) and a tubular bag-type separator should be provided.

(2) For hospitals, provide a special hospital-type separator. For hospital-type separators, provide duplex pumps each with a capacity of 0.3 gpm per operator.

(3) Since separators are high, carefully check available head room for installation.

H. Medical gas systems.

1. General.

a. Principles of design.

(1) Design all medical gas systems in accordance with the requirements of NFPA Standards 56F (Non-Flammable Medical Gas Systems) and 50 (Bulk Oxygen Systems), local authorities, local codes, and for federally financed projects, HEW standards.

(2) Medical gas systems include the following systems: Oxygen, nitrous oxide, medical compressed air, carbon dioxide, helium, nitrogen and mixtures of such gases when used for medical purposes.

(3) Make ample provision in the supply source and in the piping for anticipated future building expansion.

(4) Locate the risers with relation to the areas to be served in order to achieve minimum length branches.

(5) Arrange the risers so that patient's outlets on the floor of any wing are divided between at least two risers. A valved branch should preferably serve no more than 12 patient's outlets.

(6) A major operating suite should be served by more than one riser.

(7) Valves:

(a) Control (shutoff) valves should be ball valves and installed in valve boxes with plastic windows. Where valves are grouped together, they should be installed in a common multiple-valve box. For branches 1-1/4 in and larger, valves may have to be in separate adjacent boxes.

(b) Provide the main supply line with a shutoff valve so located as to be accessible in an emergency.

(c) Provide each riser supplied from the main line with a shutoff valve adjacent to the riser connection.

(d) Provide each lateral branch line serving patient's room outlets with a shutoff valve that controls the flow to the patient room outlets.

(e) Locate a shutoff valve outside each anesthetizing location, operating room, delivery room, etc., in each line, so located as to be readily accessible at all times for use in an emergency. Arrange these valves so that shutting off the supply of gas to any one operating room or anesthetizing location will not affect the others.

(8) The type of outlet must be as approved or required by the owner. Where multiple outlets occur, they are usually mounted on a common plate in order from left to right: oxygen, nitrous oxide, compressed air, and vacuum.

(9) Medical laboratory facilities may require in isolated locations special gases such as oxygen, hydrogen, helium, carbon dioxide, nitrogen, or argon.

(a) These requirements are usually supplied from local cylinders rather than from a central piped system.

(b) Refer to the following NFPA Standards for special requirements for these gases: 50A, Gaseous Hydrogen Systems; 56C, Laboratories in Health Related Institutions.

b. Piping.

(1) All piping should be type L copper tubing with wrought copper or cast-brass fittings with silver brazed joints, or red brass pipe with threaded cast-brass fittings.
(2) All pipe, fittings, and valves should be specially washed for medical gas use and protected from contamination thereafter.
(3) Buried piping should be adequately protected against frost, corrosion, and physical damage by installation within a pipe or conduit, and with proper cover.

c. Alarm systems.

(1) Provide alarm systems as required in NFPA Standard 56F.
(2) Locate alarm stations on the drawings.
(3) Include pressure gauges on local alarm stations.

2. Oxygen systems.

a. Principles of design.

(1) Large installations should be supplied from a bulk oxygen installation on the site. Small installations should be supplied from an oxygen cylinder manifold.

1 to 100 outlets, manifold
Over 100 outlets, bulk

(2) A rule of thumb for the preliminary sizing of the oxygen supply units (check with the supplier before final completion) is as follows:

(a) Bulk systems: Allow 500 ft^3 per bed per month plus a reserve manifold of one day's supply.
(b) Manifold systems: Allow one cylinder per bed for the main bank and double to provide for the reserve bank. 1 cylinder = 244 ft^3. Allow

space in the manifold room for the storage of spare cylinders.

(3) Oxygen supply unit: the location of the unit (bulk or manifold) should be developed between the design team, the architect, and the supplier. Consideration should be given to truck access and minimum clearances from adjacent structures, as required by NFPA Standard 50.

(4) Outlets are normally provided at least in the locations shown on pages 9-73 and 9-74 and with the indicated required flows and simultaneous-use factors. Outlets may also be required in other rooms; check the program for the exact number and locations of all outlets.

b. Pipe sizing.

(1) Calculate pipe sizing on the Lpm loads and simultaneous-use factors given above.

(2) Base pipe-size selection, in each case, on the more stringent of the following requirements:

(a) Maximum friction loss rate of 1 psi per 100 ft.

(b) Maximum friction loss to the farthest outlet of 5 psi.

(3) Minimum size for risers should be 3/4 in. The minimum size for branches should be 1/2 in.

(4) Outlet pressure should be a minimum of 50 psi.

(5) See Calculation Form P, Section 2, for sizing piping.

(6) Table 9-4 gives the friction losses in the oxygen piping. Convert the actual run of the piping to equivalent developed length by adding a fitting and valve allowance before using.

Location	Simultaneous-use factor	Volume, Lpm
First Operating room (far end of a section of piping and all individual branches to operating rooms)	100	50 per operating room[a] (two outlets per operating room)
Second Operating room (on a section of piping)	100	30 per operating room[a] (two outlets per operating room)
Each additional Operating room (on a section of piping)	100	20 per operating room[a] (two outlets per operating room)
Emergency rooms[b]	100	Same as operating room
Trauma rooms	100	Same as operating room
Delivery rooms	100	Same as operating room
Cystoscopy and Special Procedures rooms	100	Same as operating room
Recovery rooms (one outlet per bed)		20 per outlet
one to eight outlets	100	
plus nine to twelve outlets	60	
plus thirteen to sixteen outlets	50	
plus additional outlets	45	
ICU rooms (two outlets per bed)	100	20 per outlet[c]
CCU rooms (two outlets per bed)	100	20 per outlet[c]
Other spaces such as:		20 per outlet[d]
Patient rooms (Medical and Surgical) (bedside outlets)		
Sometimes one outlet per bed		
Sometimes one outlet per two beds		
Labor rooms		
Sometimes one outlet per bed		
Sometimes one outlet per two beds		
Nurseries (check program for number of outlets)		
Special Care Nursery (one outlet per incubator)		
Examination and Treatment rooms		
Operating room bed holding areas		
Surgical Preparation rooms		
Blood Donor rooms		
Anesthesia Work rooms		
Plaster (Fracture) rooms		
Dental Surgery		
Cardiac and Heart Catheterization rooms		
Deep Therapy rooms		
Inhalation Therapy rooms		
Electroencephalogram (EEG) rooms		
Electrocardiogram (ECG) rooms		

(Continued on next page)

(Continued)

Location	Simultaneous-use factor	Volume, Lpm
Electromyogram (EMG) rooms		
Fluoroscopy rooms		
High-level Radioisotope rooms		
Low-level Radiation rooms		
X-ray rooms		
Endoscopy room		

[a]Where oxygen is used to power fluidically controlled anesthesia ventilators, increase Lpm volume by 40 percent.

[b]All outlets in the emergency department (area) should have 100 percent simultaneous-use factors.

[c]Where oxygen is used to power fluidically controlled ventilators, volume should be 40 Lpm.

[d]Simultaneous-use factors for other spaces: The first outlet on the end section of piping 20 Lpm, 100 percent use factor. For additional outlets on the section of piping 10 Lpm with the following use factors:

Outlets	Percent	Lpm
1 - 3	100	
4 - 12	75	Min 45
13 - 20	50	" 115
21 - 40	33	" 125
40 and over	25	" 155

Table 9-4 Pressure Loss per 100 ft of Pipe, psi

Oxygen flow, Lpm	Nominal pipe size, in 1/2	3/4	1	1-1/4	1-1/2	2	2-1/2	3	4
50	0.04								
100	0.16								
125	0.25								
150	0.33	0.04							
175	0.48	0.06							
200	0.63	0.07							
250	0.99	0.11							
300	1.41	0.16	0.04						
400	2.51	0.29	0.07						
500	3.92	0.45	0.11						
750		1.02	0.24						
1,000		1.80	0.42	0.13	0.05				
1,250		2.81	0.66	0.21	0.09				
1,500		1.05	0.95	0.30	0.12				
2,000				0.67	0.22	0.05			
2,500				0.83	0.34	0.08			
3,000				1.19	0.49	0.11			
4,000				2.11	0.88	0.20	0.06		
5,000				3.30	1.36	0.32	0.10		
7,500					3.10	0.71	0.22	0.09	
10,000						1.27	0.40	0.16	
15,000						2.82	0.89	0.35	0.08
20,000						5.00	1.58	0.63	0.15
25,000							2.47	0.98	0.23
30,000							3.55	1.40	0.31
40,000								2.48	0.59
50,000								3.90	0.92

Source: Reprinted by permission from Building Systems Design.

3. Nitrous oxide systems.

a. Principles of design.

(1) A rule of thumb for the preliminary sizing of the inside nitrous oxide manifolds (check with the supplier before final completion): Allow one cylinder per operating, delivery, or equivalent room for the main bank and double to provide for the reserve bank. Allow space in the manifold room for the storage of spare cylinders.

(2) Nitrous oxide supply unit: Outside manifolds should be avoided due to cold-weather operational problems.

(a) The location of the unit should be

developed between the design team and the architect.

(b) Consideration should be given to truck access and minimum clearances.

(3) Nitrous oxide outlets are normally provided in the following locations and with the indicated required flows. Outlets may also be required in other rooms; check the program for the exact number and locations of all outlets.

Location	Volume, Lpm
First Operating room (far end of piping and all individual branches to operating rooms)	30 per operating room
Second Operating room (on a section of piping)	20 per operating room
Each additional Operating room (on a section of piping)	15 per room
Delivery rooms	20 per room
Emergency rooms	20 per room
Trauma rooms	20 per room
Anesthesia Work room	15 per room
Plaster (Fracture) rooms	20 per room
Endoscopy room	15 per room
Dental surgery	15 per room

b. Pipe sizing.

(1) Calculate pipe sizing on the Lpm loads given above with a simultaneous-use factor of 100 percent.

(2) Base pipe-size selection, in each case, on the more stringent of the following requirements:

(a) Maximum friction loss rate of 1 psi per 100 ft.

(b) Maximum friction loss to the farthest outlet of 5 psi.

(3) Minimum size for risers should be 3/4 in. Minimum size for branches should be 1/2 in.

(4) Outlet pressure should be a minimum of 50 psi.

(5) See Calculation Form Q, Section 2, for sizing piping.

Table 9-5 Pressure Loss per 100 ft of Pipe, psi

Nitrous oxide flow, Lpm	Nominal pipe size, in								
	1/2	3/4	1	1-1/4	1-1/2	2	2-1/2	3	4
50	0.04								
100	0.16								
125	0.25								
150	0.33	0.04							
175	0.48	0.06							
200	0.63	0.07							
250	0.99	0.11							
300	1.41	0.16	0.04						
400	2.51	0.29	0.07						
500	3.92	0.45	0.11						
750		1.02	0.24						
1,000		1.80	0.42	0.13	0.05				
1,250		2.81	0.66	0.21	0.09				
1,500		1.05	0.95	0.30	0.12				
2,000				0.67	0.22	0.05			
2,500				0.83	0.34	0.08			
3,000				1.19	0.49	0.11			
4,000				2.11	0.88	0.20	0.06		
5,000				3.30	1.36	0.32	0.10		
7,500					3.10	0.71	0.22	0.09	
10,000						1.27	0.40	0.16	
15,000						2.82	0.89	0.35	0.08
20,000						5.00	1.58	0.63	0.15
25,000							2.47	0.98	0.23
30,000							3.55	1.40	0.31
40,000								2.48	0.59
50,000								3.90	0.92

Source: Reprinted by permission from Building Systems Design.

(6) Table 9-5 gives the friction losses in the nitrous oxide piping. Convert the actual run of the piping to equivalent developed length by adding a fitting and valve allowance before using.

4. Nitrogen systems.

a. Principles of design.

(1) A rule of thumb for the preliminary sizing of the nitrogen manifolds (check with the supplier before final completion): Allow 1-1/2 cylinders per operating room requiring nitrogen, and double to provide for the reserve bank. Allow space in the manifold room for the storage of spare cylinders.

(2) Develop the location of the supply unit between the design team and the architect. Consideration should be given to truck access and minimum clearances.
(3) Provide nitrogen outlets in operating rooms for driving surgical tools when required by the program.
(4) Outlets may also be required in other rooms. Check the program for the exact number and location of all outlets.
(5) Contact the architect, owner, and/or tool manufacturer about the type of tools being considered and their volume and pressure requirements.

b. Pipe sizing.

(1) Calculate pipe sizing on the required cfm per outlet with a simultaneous-use factor of 100 percent.
(2) Base pipe-size selection, in each case, on the more stringent of the following requirements:

(a) Maximum friction loss rate of 2 psi per 100 ft
(b) Maximum friction loss to the farthest outlet of 10 psi

(3) Table 9-6 gives the friction losses in the nitrogen piping. Convert the actual run of the piping to equivalent developed length by adding a fitting and valve allowance before using.
(4) The minimum size for risers should be 3/4 in. The minimum size for branches should be 1/2 in.

c. Outlets.

(1) Outlets should be in a control box containing control valve, pressure regulator, pressure gauges, and special outlet connection.

Table 9-6 Pressure Loss (psi) per 100 ft of 160 psi Nitrogen Piping

	Nominal pipe size, in					
cfm	1/2	3/4	1	1-1/4	1-1/2	2
5	0.11	0.01				
10	0.43	0.07	0.02	0.01		
15	0.96	0.12	0.04	0.01		
20	1.70	0.26	0.07	0.02	0.01	
25	2.66	0.42	0.11	0.03	0.01	
30		0.59	0.17	0.04	0.02	
35		0.81	0.22	0.05	0.02	
40		1.06	0.29	0.07	0.03	
45		1.34	0.37	0.09	0.04	0.01
50		1.65	0.46	0.11	0.05	0.01
60		2.37	0.66	0.15	0.07	0.02
70			0.90	0.21	0.09	0.02
80			1.17	0.27	0.12	0.03
90			1.48	0.34	0.15	0.04
100			1.83	0.43	0.19	0.05
110			2.21	0.51	0.23	0.06
120				0.62	0.27	0.07
130				0.72	0.32	0.09
140				0.83	0.37	0.10
150				0.96	0.42	0.11

I. Central soap systems.

1. Principles of design.

a. Central soap systems are usually provided only when requested by the owner and/or architect. (You can suggest them where suitable.) Where only individual soap dispensers are being used, check with the architect whether they are to be provided by the plumbing contractor or the general contractor.
b. Provide batteries of three or more lavatories with central soap-dispensing systems.
c. Provide other lavatories, except patients' private lavatories, with individual soap dispensers.
d. Provide batteries of three or more gang shower heads with central soap-dispensing systems.
e. Provide other shower heads with individual soap dispensers.
f. Where batteries of lavatories are over one another on two or more floors, they should be fed from a tank on the uppermost floor with

valved branches serving each floor. Not more than four floors should be fed from one tank (maximum 40 ft from tank to lowest outlet).

g. Wherever possible, locate tanks in janitor's closets. When located in toilet rooms, place in men's toilet rooms in preference to women's toilets whenever possible. Tanks should be surface-mounted or recessed as per architect's preference.
h. Where stacked toilet rooms exceed four floors (the limit of a single tank), a master tank can be located in a machinery space directly above the local tanks and piped to manually fill the local tanks. Check with manufacturer about suggested sizing.
i. Run piping concealed, except for branches, to surface-mounted tanks and basin-mounted dispensing valves.
j. Put floor control valves, unless in janitor's closets, in or behind the wall with satin-finish stainless-steel access doors.

2. Piping.

a. Exposed piping should be chromium-plated standard-weight brass pipe with threaded standard-weight cast brass fittings.
b. Concealed piping should be standard-weight black steel pipe with threaded standard-weight malleable iron fittings.
c. All piping should be 3/8 in.

3. Equipment.

a. Local tanks.

(1) Recessed or exposed satin finish stainless steel of sizes as follows:

1 gal,	up to 4 outlets
2 gals,	5 to 8 outlets
2-1/2 gals,	5 to 12 outlets
3 gals,	9 to 15 outlets
5 gals,	13 to 25 outlets

(2) 2-gal size also available in baked white enameled steel.

b. Master tanks.

(1) Round stainless steel in 25-, 50-, 100- and 250-gal sizes.

4. Outlets.

a. Lavatory outlets should be liquid lather type, either wall or basin-mounted type as desired.

5. Individual dispensers.

a. Individual lavatory dispensers should be chromium-plated liquid lather as follows:

(1) Basin-mounted type.
(2) Surface-mounted type.
(3) Recessed type.
(4) Built as part of a shelf.

b. Individual shower dispensers should be chromium-plated liquid type with vertical valve.
c. Shower outlets should be liquid type with vertical valve.

Section **10**

Interior Fire Protection

Design of interior fire-protection systems, typically sprinklers, standpipes, and chemical systems, is often a task of the plumbing designer. This section provides detailed guidance for the design of such systems.

Sprinklers are the most common form of fire-suppression systems in industrial structures. They are discussed in detail here. Likewise, systems using any of the various chemical suppressants are also considered.

Close compliance with prevailing codes is necessary in the design of any fire-protection system. These codes are discussed in this section, also.

FIRE-SUPPRESSION SYSTEMS

This section describes systems which extinguish or control fires.

Water Supply

The most common type of fire-suppression systems rely on water. Therefore it is essential that adequate supplies of water be provided and maintained. A thorough discussion of water-supply systems is presented in Sections 3 and 4 of this reference guide.

The plant water-supply system or nearby public water supply will be the primary fire-suppression system used by the plant fire brigade or public fire department. Water must be provided in quantity and pressure sufficient for supplying automatic sprinkler systems and fire hoses, in addition to normal plant requirements. When the public water supply is inadequate for plant protection, supplemental private supplies are necessary.

Pipe Networks

The minimum recommended pipe size for fire protection is 6 in; the pipe network is looped in a grid pattern, and no leg is more than 600 ft long.

It is usually cost-effective to use larger pipe sizes since the installation costs are relatively the same (Table 10-1).

TABLE 10-1 Comparison of Pipe Capacity

Size of pipe, in	Relative capacity
6	1.0
8	2.1
10	3.8
12	6.2
14	9.3
16	13.2

Source: NFPA, Quincy, Massachusetts.

Wherever possible pipe networks should be arranged in loops to provide water flow from two directions when necessary (Fig. 10-1).

A pipe 8 in or larger is necessary to supply:

- More than one hydrant on a dead-end main
- A dead-end main exceeding 500 ft
- Two hydrants on a loop exceeding 1500 ft
- Three hydrants on a loop exceeding 1000 ft
- Four or more hydrants
- Hydrants with three or four outlets
- Water at low pressure

Fire Hydrants

Fire hydrants are provided on public mains to allow the fire department to draw water with mobile pumpers to supply sprinkler and standpipe systems, or hose streams. Fire hydrants are provided on private mains to allow the fire brigade or fire department to supply hose streams and where the supply is adequate to support sprinkler and standpipe systems with mobile pumpers.

Hydrants are available in wet-barrel (California, Fig. 10-2) and dry-barrel (base valve, Fig. 10-3) types. Base-valve hydrants are necessary where there is any chance of freezing.

Hydrants on plant pipe networks should be located every 250 ft and about 50 ft from the buildings protected. They must be protected from damage by vehicles or machinery.

The available water flow for fire suppression is determined by flow testing the hydrant system. The water flow at 20 psig is calculated since this is the minimum pressure required by fire department pumpers.

Valves in pipe lines supplying fire-protection water are generally required to be indicating valves. These include underground gate valves with indicator post, underground butterfly valves with indicator post, and outside screw and yoke (OS&Y) gate valves.

Incorrectly shut valves have been the primary cause of sprinkler systems failing to control fires.

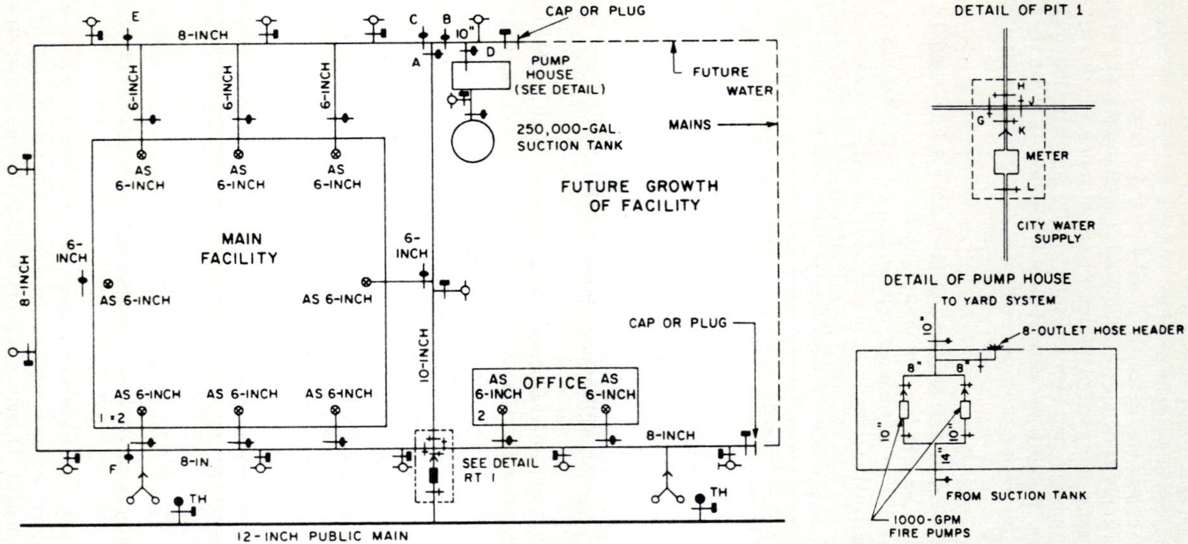

Figure 10-1 Water-pipe network. *(From* Fire Protection Handbook, *14th ed., G. P. McKinnon (ed.), NFPA, Quincy, Massachusetts, 1976.)*

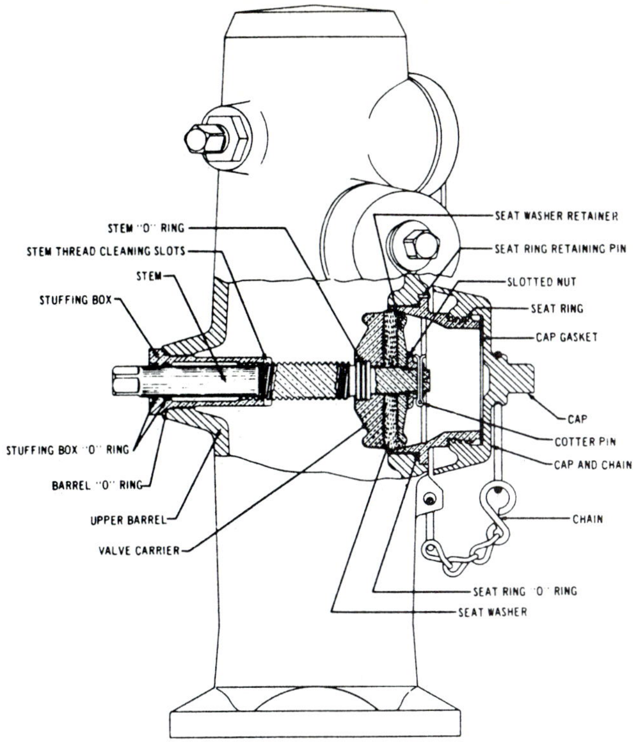

Figure 10-2 Wet-barrel hydrant. *(Mueller Co.)*

Fire Pumps

Fire pumps are essentially the same as water-supply pumps discussed in Section 3. Additional considerations required for fire pumps include:

- Use of equipment approved for fire pumps
- Use of approved accessories
- Adequate capacity to meet fire-flow demands
- Automatic operation
- Safe location for uninterrupted service
- Annual testing
- Maintenance
- Selection of fire pump driver based on reliability, adequacy, economy, and safety of power source

Sprinkler Systems

A sprinkler system, for fire-protection purposes, is an integrated system of underground and overhead piping designed in accordance with fire-protection engineering standards. The installation includes a water supply, such as a gravity tank, fire pump, reservoir or pressure tank and/or connection by underground piping to a city main. The portion of the sprinkler system above ground is a network of specially sized or hydraulically designed piping installed in a building, structure, or area, generally overhead, and to which sprinklers are connected in a systematic pattern. The system includes a controlling valve and a device for actuating an alarm when the system is in operation. The system is usually activated by heat from a fire and discharges water over the fire area.

Wet-Pipe Systems

These are under water pressure at all times; water will be discharged immediately when an automatic sprinkler operates (Fig. 10-4).

Water flowing from the wet-pipe sprinkler system actuates an alarm valve that gives off a signal. Figure 10-4 illustrates the total concept of the wet-pipe automatic sprinkler system.

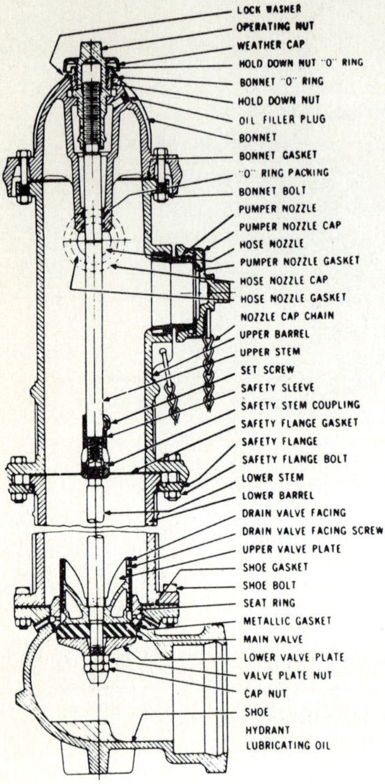

Figure 10-3 Dry-barrel hydrant. *(Mueller Co.)*

The essential features of wet-pipe sprinkler systems, which represent about 75 percent of sprinkler installations, include provisions for water supplies, piping, and location and spacing of sprinklers. This system is generally used wherever there is no danger that the water in the pipes will freeze and wherever there are no special conditions requiring one of the other systems. Inspection of the wet-pipe sprinkler system at regular intervals is essential, and weekly inspection of all water-control valves and alarm-control valves is recommended.

Where subject to temperatures below freezing, the ordinary wet-pipe system cannot be used. There are two recognized methods of maintaining automatic sprinkler protection in such locations: (1) through the use of systems where water enters the sprinkler piping only after operation of a control valve (dry-pipe, pre-action, etc.) and (2) by the use of antifreeze solution in a portion of the wet-pipe system.

Regular Dry-Pipe Systems

In locations where there is danger of freezing, it is the usual practice to install a dry-pipe system. In regular dry-pipe systems, the sprinkler piping contains air or nitrogen under pressure instead of water, and admission of the water is controlled by a dry-pipe valve. When a sprinkler is opened by heat from a fire, the pressure is reduced, a dry-pipe valve is opened by water pressure, and water flows out of any opened sprinklers.

Pre-Action Systems

Systems in which the air in the piping may or may not be under pressure are called pre-action systems. These systems are designed primarily to protect properties on which the danger of water damage from broken sprinklers or piping could be serious. The water-supply valve is actuated independently of the opening of the sprinkler heads by an automatic fire-detection system; the valve is opened sooner than with the dry-pipe system, and the alarm is given when the valve is opened.

The pre-action system has several advantages over a dry-pipe system. The valve is opened sooner because the fire detectors have less thermal lag than sprinklers. The detection system also automatically rings an alarm. Fire and water damage is decreased because water is on the fire sooner, and the alarm is given when the valve is opened. Sprinkler piping is normally dry; thus, pre-action systems are nonfreezing and applicable to dry-pipe service.

Deluge Systems

Deluge systems are used for areas of extra occupancies. All sprinkler heads are open at all times so that when the water comes on, the entire area is flooded; when heat from a

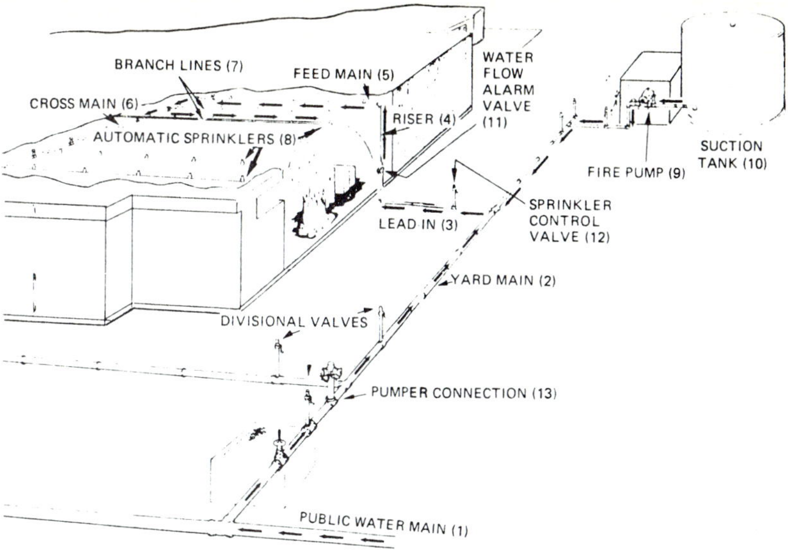

Figure 10-4 Sprinkler system. *(Factory Mutual System.)*

fire actuates the fire-detecting device, water flows to and is discharged from all sprinklers on the piping system, thus "deluging" the protected areas. These systems are often used in airplane hangars and in areas where flammable liquids are handled or stored.

By using sensitive thermostatic controls operating on the rate-of-rise or fixed-temperature principle, or controls designed for individual hazards, it is possible to apply water to a fire more quickly than with systems in which operation depends on opening of sprinklers only as the fire spreads.

Sprinkler Heads

Sprinkler heads are designed with temperature ratings ranging from 135°F (57°C) to as high as 500°F (260°C). Ratings of 165°F (74°C) are usual for use in buildings that are maintained at normal, constant temperatures.

The location and spacing of sprinkler heads depends on the degree of hazard and type of construction. The "Standard for the Installation of Sprinkler Systems," NFPA 13, provides detailed design and maintenance specifications (Table 10-2).

Standpipes

The four generally recognized standpipe system concepts are described in (Table 10-3):

1. A wet-standpipe system, having supply valve open and water pressure maintained at all times. This is the most desirable type of system.
2. A dry-standpipe system arranged to admit water to the system through manual operation of approved remote-control devices located at each hose station. The water-supply control mechanism introduces an inherent reliability factor that must be considered.
3. A dry-standpipe system in an unheated building. The system should be arranged to admit water automatically by means of a dry-pipe valve or other approved device. The depletion of system air at the time of use introduces a delay in the application of water to the fire and increases the level of competency required to control the pressurized hose and nozzle assembly during the charging period.

TABLE 10-2 Summary of Spacing Rules

Type of construction	Maximum distance of deflectors below ceiling, in			
	In bays*		Under beams	
	Comb.	Noncomb.	Comb.	Noncomb.
Smooth ceiling	10	12	14	16
Beam and girder	16	16	20	20
Panel up to 300 ft^2	18	18	22	22
Bar joists	10	12	—	—
Open wood joists—center 3 ft or less	6	—	—	—

Minimum below ceiling is 1 in.
Minimum below beams 1 in, maximum 4 in. Do not exceed maximum below ceiling.

Maximum coverage per sprinkler	
Light hazard	200 ft^2 smooth ceiling and beam and girder construction
	130 ft^2 open wood joist
	168 ft^2 all other types of construction
Ordinary hazard	130 ft^2 all types of construction except
	100 ft^2 high piled storage
Extra hazard	90 ft^2 all types of construction

Direction of lines: Either direction to facilitate hanging except: across beams for beams on girders 3 ft to 7½ ft on centers and across joists for wood joists (open or sheathed) and bar joists (through or under)

Maximum spacing between lines and sprinklers	
Light and ordinary hazard	15 ft except 12 ft for high piled storage
Extra hazard	12 ft

*Not more than 4 in below beams where lines run across beams.
Source: NFPA 13, "Installation of Sprinkler Systems," Quincy, Massachusetts.

4. A dry-standpipe system having no permanent water supply. This type would be used for reducing the time required for fire departments to put hose lines into action on upper floors of tall buildings. This type of system might also be used in buildings during construction, where allowed in lieu of the wet standpipe in unheated areas.

Water-Spray Fixed Systems

Water-spray fixed systems are generally used to protect flammable liquid and gas tankage; piping and equipment; electrical equipment such as transformers, oil switches, and rotating electrical machinery; and openings in firewalls and floors through which conveyors pass. The type of water spray required for any particular hazard depends, of course, upon the nature of the hazard and the purpose for which the protection is provided.

NFPA 15, "Standard for Water Spray Fixed Systems for Fire Protection," calls for piping, valves, pressure gauges, and detection systems of an approved type. The spray nozzles generally used in these systems are open, and the pipes, especially outdoor ones that are subject to freezing temperatures, are usually dry.

Foam Extinguishing Systems

Foam extinguishing systems have been used extensively for many years, especially in the petrochemical industry, for the extinguishment of flammable liquid fires. The principal

TABLE 10-3 Summary of National Fire Protection Association Standpipe Standards*

Type	Intended use	Size hose and distribution	Minimum size pipe	Minimum water supply
Class I	Heavy streams	2½-in connections	4 in up to 100 ft	500 gal/min 1st standpipe
	Fire department	All portions of each story or section within 30 ft of nozzle with 100 ft of hose	6 in above 100 ft	250 gal/min each additional 2500 gal/min maximum)
	Trained personnel			
	Advanced stages of fire		(275 ft maximum unless pressure regulated)	30-min duration 65 lb/in^2 at top outlet with 500 gal/min flow
Class II	Small streams	1½-in connections (distribution same as class I)	2 in up to 50 ft	100 gal/min per building
	Building occupants		2½ in above 50 ft	30-min duration
	Incipient fire			65 lb/in^2 at top outlet with 100 gal/min flowing
Class III	Both of above	Same as class I with added 1½-in outlets or 1½-in adapters and 1½-in hose.	Same as class I	Same as class I

*From NFPA 14, "Standard for the Installation of Standpipe and Hose Systems," Quincy, Massachusetts.

kinds of foam are chemical and mechanical (determined by how they are generated), and these classes are further subdivided.

Special compatible foam concentrates result in the generation of a foam that does not break down as readily as ordinary foam when mixed with dry chemical. Other special foams are available for application on fires in alcohols, esters, ketones, and ethers (called water-soluble or polar liquids). This concentrate produces a foam that does not deteriorate like ordinary foam when in contact with water-miscible solvents.

Carbon Dioxide Systems

Carbon dioxide is a noncombustible gas that has been effectively used to extinguish certain types of fires. It acts to reduce the oxygen in the fire area to a point where it will no longer support combustion (Table 10-4). Because carbon dioxide is stored under pressure, it can readily be discharged from its cylinder or extinguisher. Carbon dioxide is inert and will not conduct electricity. It can be used safely on energized electric equipment fires without causing damage to the equipment.

Because carbon dioxide does little or no damage to equipment or materials with which it comes in contact, it is very useful for protection of rooms with contents of high value and contents subject to water damage. Typical of such occupancies are fur vaults, record storage rooms, computer rooms, and rooms housing live electric equipment. Carbon dioxide is also widely used for extinguishing flammable liquid fires because the carbon dioxide gas rapidly spreads above the surface of the burning liquid and shuts off the oxygen.

Halogenated Agents and Systems

One of the earliest halons used for fire extinguishing was carbon tetrachloride. Concern about the toxic properties of carbon tetrachloride began to manifest itself about 1960, and the *NFPA and the extinguisher-testing laboratories discontinued any recommendation of carbon tetrachloride as a fire extinguishing agent.*

TABLE 10-4 Minimum Carbon Dioxide Concentrations for Extinguishment*

Material	Theoretical min. CO_2 concentration, (%)
Acetylene	55
Acetone	26†
Benzol, benzene	31
Butadiene	34
Butane	28
Carbon disulfide	55
Carbon monoxide	53
Coal gas or natural gas	31†
Cyclopropane	31
Dowtherm	38†
Ethane	33
Ethyl ether	38†
Ethyl alcohol	36
Ethylene	41
Ethylene dichloride	21
Ethylene oxide	44
Gasoline	28
Hexane	29
Hydrogen	62
Isobutane	30†
Kerosene	28
Methane	25
Methyl alcohol	26
Pentane	29
Propane	30
Propylene	30
Quench, lubricating oils	28

*From Gordon P. McKinnon (ed.) *Fire Protection Handbook,* NFPA, Quincy, Massachusetts, 1976.

†The theoretical minimum extinguishing concentrations in air for the above materials were obtained from Bulletin 503, U.S. Bureau of Mines, Washington, 1952. Those marked † were calculated from accepted residual oxygen values.

A halon is a hydrocarbon (hydrogen and carbon) in which some of the hydrogen atoms have been replaced by such elements as bromine, chlorine, or fluorine, or by combinations of these (Table 10-5). A number of halons are toxic, thus making them undesirable for general use; two of them, Halon 1301 and Halon 1211, have acceptable levels of toxicity and excellent flame extinguishment properties.

Halon 1211 and Halon 1301 are the only two agents recognized by the NFPA Technical Committee on Halogenated Fire Extinguishing Agent Systems. Both Halon 1211 and 1301 are widely used for protection of electric equipment (both are nonconductors of electricity), airplane engines, and computer rooms. As both of these halons rapidly vaporize, they leave little corrosive or abrasive residue to clean up and do not interfere as much with visibility during fire fighting as foam or carbon dioxide. These halons are used for hand extinguishers and fixed systems.

Dry-Chemical Extinguishing Systems

Dry-chemical extinguishing agents consist of finely divided powders that effectively extinguish a fire when applied to the fire by portable extinguishers, hose lines, or fixed systems. The original dry powder was sodium bicarbonate (ordinary baking soda). Potassium bicarbonate and other chemical powders, with additives to make the powders free flowing and more moisture resistant, are now in use. Dry chemical has been found to be

TABLE 10-5 Some Physical Properties of the Common Halogenated Fire Extinguishing Agents

Agent	Chemical formula	Halon no.	Type of agent	Approx. boiling point, °F	Approx. freezing point, °F	Specific gravity of liquid at 68°F (water = 1)	Approx. critical temp., °F	Estimated pressure, psig		Latent heat of vaporization, cal/g water = 540 cal/g CO_2 = 138 cal/g
								At 130°F	At critical temp.	
Carbon tetrachloride	CCl_4	104	Liquid	170	−8	1.595	—	—	—	46
Methyl bromide	CH_3Br	1001	Liquid	40	−135	1.73	—	—	—	62
Bromochloromethane	$BrCH_2Cl$	1011	Liquid	151	−124	1.93	—	—	—	—
Dibromodifluoromethane	Br_2CF_2	1202	Liquid	76	−223	2.28	389	23	585	29
Bromochlorodifluoromethane	$BrCClF_2$	1211	Liquefied gas*	25	−257	1.83	309	75	580	32
Bromotrifluoromethane	$BrCF_3$	1301	Liquefied gas	−72	−270	1.57	153	435	560	28
Dibromotetrafluoroethane	BrF_2CCBrF_2	2402	Liquid	117	−167	2.17	—	3.8	—	25

*May be kept as a liquid at reduced temperatures.

Source: NFPA, Quincy, Massachusetts.

an effective extinguishing agent for fires in flammable liquids and in certain types of ordinary combustibles and electric equipment, depending upon the type of dry chemical used.

Dry-chemical extinguishing systems are used to protect flammable-liquid storage rooms, dip tanks, kitchen range hoods, deep-fat fryers, and similar hazardous areas and appliances. Because dry chemical is nonconductive, these systems are useful in the protection of oil-filled transformers and circuit breakers. Dry-chemical systems are not recommended for telephone-switchboard or computer protection. Dry-chemical hose-line systems are used in crash trucks and for the protection of aircraft hangars. Dry chemicals are used in portable fire extinguishers.

Combustible-Metal Extinguishing Systems

A number of metals and metal powders found in industrial situations and in transport will burn. Some metals burn when heated to high temperatures by friction or exposure to external heat. Others burn from contact with moisture or in reaction with other materials. These metals and metal powders require special extinguishing agents and special fire-fighting techniques. Some result in explosions and very high temperatures, and some react violently with water. Still others give off toxic fumes when burning.

Some combustible metal extinguishing agents' success in handling metal fires has led to the terms *approved extinguishing powder* and *dry powder.* Such terms have been accepted in describing extinguishing agents for metal fires, and should not be confused with the name *dry chemical,* which normally applies to an agent suitable for use on flammable-liquid and live electric equipment fires. Graphite powder, talc, and sand have all been used to smother metal fires.

Portable Fire Extinguishers

Portable fire extinguishers are required in most plants by local, state, and federal regulations and insurance requirements. Where there are trained personnel available to use the proper extinguisher on a small incipient fire, extinguishers may prove useful in preventing a larger, more devastating fire.

The limitations of extinguishers, personal exposure to fire and smoke, capacity, range, selectivity, and availability necessitate that training be provided if they are expected to be effective. Use of extinguishers should be simultaneous with notification of the fire brigade or department.

Types of Portable Fire Extinguishers

The kind and number of extinguishers needed for particular types of fires are specified in NFPA 10, "Standard for Portable Fire Extinguishers." The most common types of extinguishers in use are the pressurized water, carbon dioxide, and multipurpose dry chemical. Other extinguishers commonly used are water pump tanks, liquefied gases such as Halon 1211, and combustible-metal-type dry powder.

Application of Portable Fire Extinguishers

NFPA 10, "Standard for Portable Fire Extinguishers," classifies fires in four ways:

Class A. Fires involving ordinary combustible materials (wood, cloth, paper, rubber, and many plastics) requiring the heat-absorbing (cooling) effects of water, water solutions, or the coating effects of certain dry chemicals that retard combustion (Table 10-6).

Class B. Fires involving flammable or combustible liquids, flammable gases, greases, and similar materials where extinguishment is most readily secured by excluding air (oxygen), inhibiting the release of combustible vapors, or interrupting the combustion chain reaction (Table 10-7).

Class C. Fires involving live electric equipment where safety to the operator requires the use of electrically nonconductive extinguishing agents. (*Note:* When electric equipment is de-energized, the use of Class A or B extinguishers may be indicated.)

TABLE 10-6 Fire Extinguisher Size and Placement for Class A Hazards

Basic minimum extinguisher rating for area specified	Maximum travel distances to extinguishers, ft	Areas to be protected per extinguisher: Light hazard occupancy, ft^2	Ordinary hazard occupancy, ft^2	Extra hazard occupancy, ft^2
1-A	75	3,000	†	†
2-A	75	6,000	3,000	†
3-A	75	9,000	4,500	3,000
4-A	75	11,250	6,000	4,000
6-A	75	11,250	9,000	6,000
10-A	75	11,250*	11,250*	9,000
20-A	75	11,250*	11,250*	11,250*
40-A	75	11,250*	11,250*	11,250*

*11,250 ft^2 is considered a practical limit.

†Protection requirements may be fulfilled by several extinguishers of the minimum specified rating with the approval of the authority having jurisdiction.

Source: NFPA, Quincy, Massachusetts.

TABLE 10-7 Fire Extinguisher Size and Placement for Class B Hazard Excluding Protection of Deep Layer Flammable Liquid Tanks (For Extinguishers Labeled After June 1, 1969)

Type of hazard	Basic minimum extinguisher rating	Maximum travel distance to extinguishers, ft
Light	5-B	30
	10-B	50
Ordinary	10-B	30
	20-B	50
Extra	20-B	30
	40-B	50

Source: NFPA, Quincy, Massachusetts.

Class D. Fires involving certain combustible metals (such as magnesium, titanium, zirconium, sodium, potassium, etc.) requiring a heat-absorbing extinguishing medium not reactive with the burning metals.

Section **11**

Sewage and Drainage Pumps and Drains

Where a sewer is higher than the lowest point in a building or other part of a property served by a plumbing system, it may be necessary to use a pump to raise effluent to the sewer level. This section discusses the various types of pumps used to accomplish this task.

The plumbing designer faced with the collection of plumbing wastes at a level lower than the sewer serving the building or area, will be required to choose a way to raise the fluid to sewer level. Knowledge of the types of pumps available, their characteristics, relative costs, and installation requirements is therefore necessary. This section discusses many of these aspects of the design of pumps for plumbing systems.

11-1. Needs for Sewage and Drainage Pumps. It is not an unusual condition to find either the sanitary sewer or the storm sewer, or both, at a higher elevation than the lowest point within the building or on the property to be drained. In large cities the construction of subbasements below the elevation of city sewers is customary. Typical plumbing for such conditions is shown in Fig. 11-1. A pump is required to lift into the building drain or into the building sewer only that portion of the sewage or drainage originating below the elevation necessary for its flow into the sewer by gravity.

11-2. Types of Sewage and Drainage Pump. Three types of pump are commonly used for ejecting sewage or drainage from a premises: centrifugal, as shown in Fig. 11-2; air displacement, as shown in Fig. 11-3, and water or steam ejector. Piston, plunger, or other displacement pumps are not suitable for such service because of the solid particles or grit in sewage and in much of the surface drainage to be removed.

Sewage and drainage pumps in buildings are usually controlled automatically, the control being by means of a float that sets the pump into motion when the sump is full and stops the pump when the sump is empty. Apparatus for ejecting sewage should be installed in duplicate so that when one pump requires maintenance attention, the other can be in operation.

11-3. Capacities of Sumps and Pumps. The required capacity of the pump is affected by the rates of flow of drainage and by the size of the sump. Where there is no sump, the capacity of the pump or pumps must equal or exceed the maximum rate of drainage flow. Where the sump is large enough to permit the pump to operate continuously, the capacity of the pump must equal or exceed only the *average* rate of drainage flow. Since information on fluctuations in drainage flow rates is usually inadequate and the capacity of a sump is limited to a maximum retention period of 12 hr,[1] it is not practicable to select a pump that will run continuously. The rate of flow on which the capacity of the pump is based should equal the sum of all the fixture units draining into it with allowance

[1] "National Plumbing Code," Sec. 11.7.2.

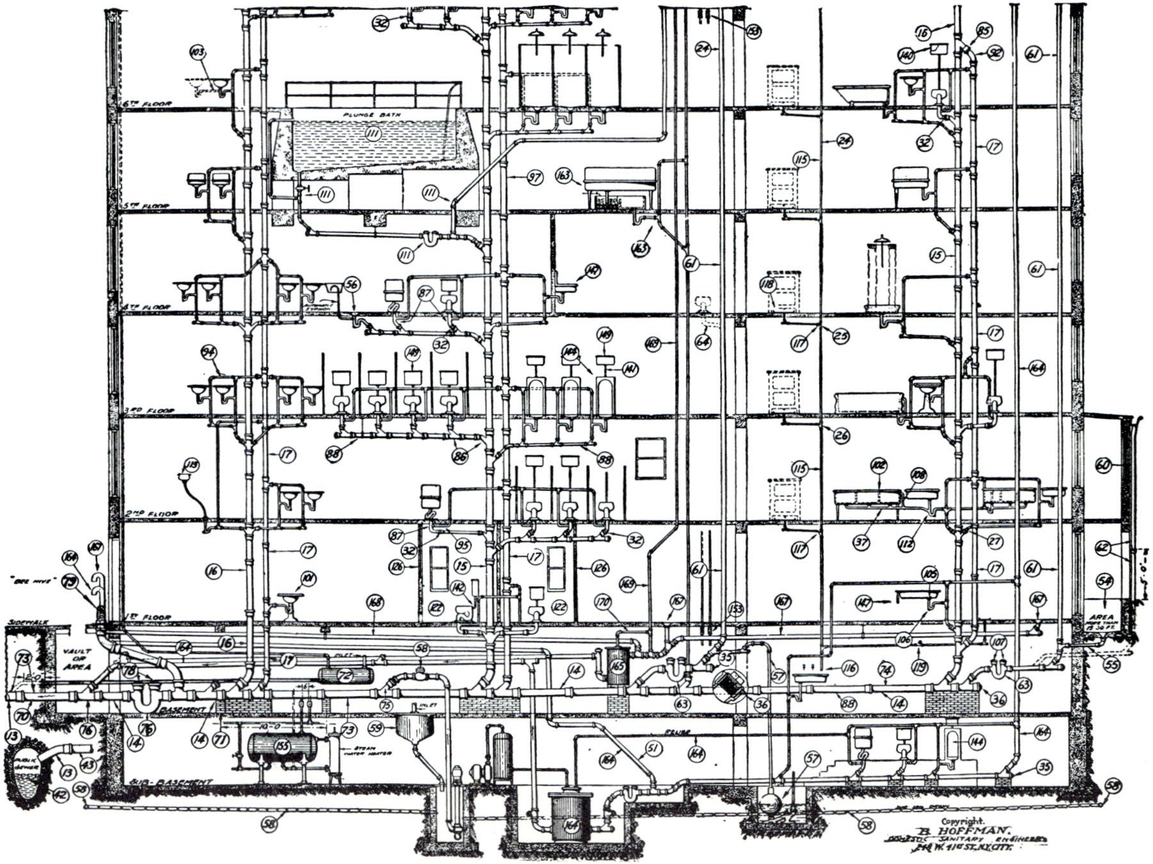

FIG. 11-1. Requirements of the New York City Plumbing Code. (*Published by permission of B. Hoffman.*)

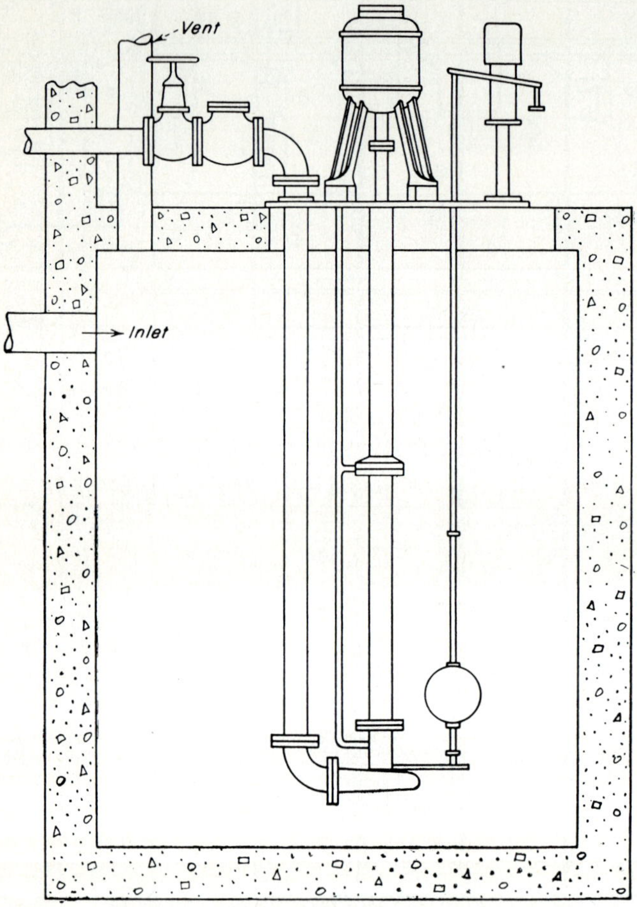

FIG. 11-2. Centrifugal pump in a wet well.

for the nonsimultaneous use of fixtures as discussed in the Code and as shown in Fig. 11-16. For example, if all fixtures discharging simultaneously delivered 1,000 gpm, the actual flow on which the capacity of the pump should be based, according to Fig. 11-16, is 210 gpm. The head against which the pump discharges is the static lift plus friction and velocity heads.

In the ordinary installation in the basement of a building where an overflow of sewage or drainage might be disastrous, the pump should have

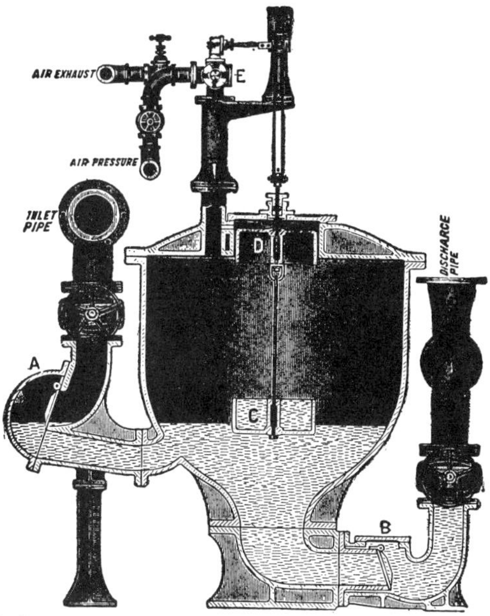

FIG. 11-3. Compressed-air sewage ejector. (*Yeomans Brothers Co.*)

a rated capacity equal to or greater than the estimated maximum rate of flow and the sump should be of such a size that it will require about 15 to 60 min to be filled under the maximum rate of flow, the longer period of storage being used for smaller installations. Where automatic control is not used, the sump should fill in not more than 12 hr and the sump should be empty when the pump stops. Pumps should be installed in duplicate in all but the smallest establishments, each pump having sufficient capacity to carry the maximum load. The "National Plumbing Code" states (Sec. 11.7.5) that duplex pumping equipment must be provided when the discharge is more than six water closets.

11-4. Automatic Sump Pumps. The operation of sump pumps for the removal of drainage from buildings should be automatic. A common type of automatic float-controlled centrifugal pump for a residential installation would be designed so that the pump goes into operation when the rising float in the basin actuates the starting switch on the motor. The power circuit is broken when the float descends to the bottom of the basin. A check valve is necessary, and a gate valve should be provided on the

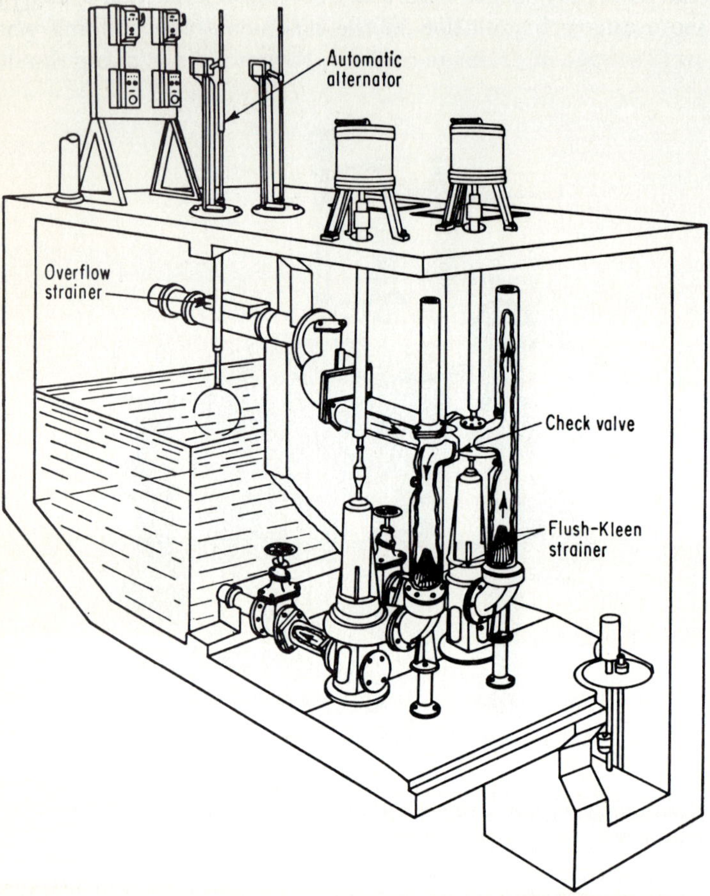

FIG. 11-4. Flush-Kleen pumps. When the pump on the right is discharging, sewage flows from the sewer through the discharge pipe of the pump on the left and into the wet well. When the wet well is empty, the pump on the right stops and the wet well refills. Then the pump on the left starts, closing the check valve on its discharge pipe, diverting sewage to the pump on the right, and throwing sewage from the wet well through the screen in its discharge pipe, which is thus cleaned. The discharge enters the building sewer or other conduit or discharge point. (*Chicago Pump Co.*)

discharge line. The check valve prevents backflow of sewage, and the gate valve, located between the check valve and the sewer, can be closed when repairs are necessary.

Sump pumps can be so arranged that when the pump is not operating, drainage flows through the screen on the *discharge* line of the pump, through the pump, and from the suction line into the sump. When the sump fills, the pump goes into operation automatically, closing a check valve on the drainage line. The direction of flow through the pump is reversed, and the screen on the discharge line of the pump is automatically flushed by the discharge from the pump. An arrangement of two such

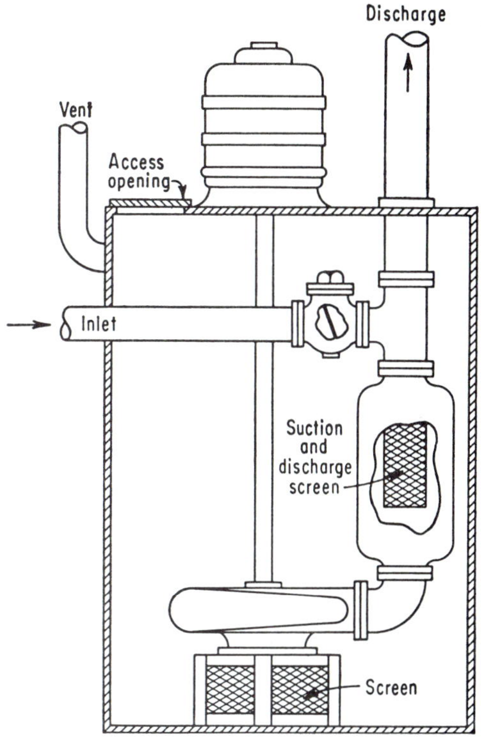

FIG. 11-5. Single, self-cleansing sump pump.

pumps which operate alternately is shown in Fig. 11-4, and a single pump of the type is shown in Fig. 11-5.

The sump pit can be constructed of concrete, tile, or metal, somewhat as shown in Fig. 11-6. The sump pit should be vented to the air outside the building to avoid the creation of nonatmospheric pressures in the sump and to remove odors from the building.

11-5. Air-displacement Pumps. The air-displacement sewage pump or compressed-air sewage ejector, such as is shown in Fig. 11-1, is often used for pumping sewage from buildings. Among its advantages for this service may be included the following: It is enclosed, thus preventing the escape of sewer air into the building. Its operation can be made completely automatic so that it goes into service only when there is need. The few moving parts in contact with sewage require little or no attention and no lubrication, and they are not easily clogged. Under normal conditions it will function for long periods of time without attention. A disadvantage of this type of pump is the need for an air compressor.

The pump shown in Fig. 11-3 operates as follows: Sewage enters the reservoir through the inlet pipe at the left, the air displaced being slowly expelled through the air exhaust, which is open to the air through a vent pipe. The rising sewage lifts the float D, which opens the valve in the

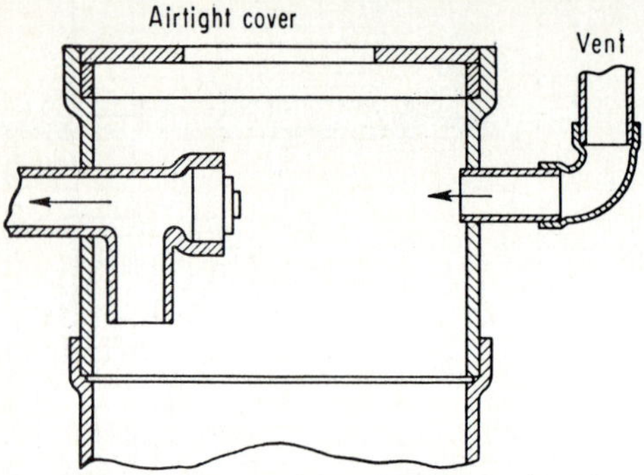

FIG. 11-6. Catch basin.

pipe above the reservoir when the reservoir fills. The opening of the valve admits compressed air to the reservoir. The air pressure closes the air exhaust and the inlet valve at the left and ejects the sewage through the discharge valve B and pipe at the right. As the chamber C drops with the descending sewage, it shuts off the air supply and opens the air exhaust through the small pipe at the top center. Sewage is prevented from flowing back into the reservoir by the check valve in the discharge pipe. The air vent from the compressor should be carried to the open air as a vent terminal.

The capacities of air compressors are stated in Table 11-1. These so proportioned and the air compressors so designed that there will always be a sufficient volume of air at a pressure sufficient to raise all the sewage or water in the receiving sump at least 50 per cent higher than the maximum lift required.

The capacities of air compressors are stated in Table 11-1. These pumps are available in capacities from 50 to 1,000 gpm. The capacity of the air compressor required is dependent on its efficiency, the height of lift, and the quantity of sewage to be lifted. The following formula

[1] See *ibid.*, Sec. 11.7.7.

will give the approximate capacity of the air compressor required:

$$A = \frac{S(34 + h)}{255}$$

where A = free air per minute drawn in by air compressor, i.e., capacity of compressor, cu ft

h = lift against which sewage is to be discharged, ft

S = rate of flow of sewage during discharge, gpm

It is customary to place an air receiver between the compressor and the pump to permit the use of a smaller compressor which will operate for a longer period of time. The compressor may be driven by an electric motor or hydraulically by the building water supply. The water used

TABLE 11-1. APPROXIMATE CAPACITIES OF AIR COMPRESSORS REQUIRED TO OPERATE AIR-DISPLACEMENT PUMPS

Water pressure, psi	Cu ft free air per gal water	Water pressure, psi	Cu ft free air per gal water	Water pressure, psi	Cu ft free air per gal water	Water pressure, psi	Cu ft free air per gal water	Water pressure, psi	Cu ft free air per gal water
5	0.179	40	0.497	70	0.815	100	1.043	130	1.316
10	0.224	50	0.588	80	0.861	110	1.134	140	1.407
20	0.315	60	0.679	90	0.952	120	1.225	150	1.498
30	0.406								

by the hydraulically driven air compressor can be made available for other uses.

11-6. Centrifugal Pumps. Centrifugal pumps are frequently used for pumping drainage water, other than sanitary sewage, from an "open" sump. A typical pump installation is shown in Fig. 11-2. Because the sump is "open," that is, the air in the sump is open to the atmosphere in the building, such an installation is not suitable for pumping sanitary sewage from a building unless the pump is placed outside. To remove sewage from a building, the centrifugal pump and sump must be totally enclosed or the sump must not be in the building. If sanitary sewage is to be pumped, a special type of nonclog impeller is required unless a self-cleaning type of screen is used on the combined inlet-discharge pipe, as explained in Sec. 11-4 and as shown in Figs. 11-2 and 11-3.

11-7. Water-ejector Sump Pump. A drainage pump operated by a water ejector is illustrated in Fig. 11-5. The ejector must not be connected to a water supply used for any domestic purpose. The ejector operates because the velocity of water passing through the throat of the jet is retarded by the expansion of the stream beyond the jet. The stream expands because of the increase in size of the channel beyond the throat. The expansion of the jet creates a vacuum which causes suction on the inlet pipe. The velocity energy of the water is converted into pressure energy which is available for lifting water from the sump. Such a pump is simple and reliable and requires little or no attention in operation and no lubrication, and the power for its operation can be obtained from the public water supply. For this reason it is particularly adapted to use in small buildings where the mechanical equipment must be of the simplest kind and where the cost of water is low. The capacities of Penberthy hydraulic ejectors are given in Table 11-2. The discharge lift of the pump is limited to about 12 to 18 ft, and the discharge line should be arranged to prevent the backflow of air. This can be done by submerging the end of the discharge pipe or by other means. The operation of the pump requires about 4 to 5 psi pressure on the power water-supply line for each foot of lift and about 1 gal of supply water for each gallon of water discharged for lifts up to about 8 ft. In order to obtain the best efficiency it is recommended that the ejector be set so that all the lift is in the suction pipe and that the total lift is less than 8 ft.

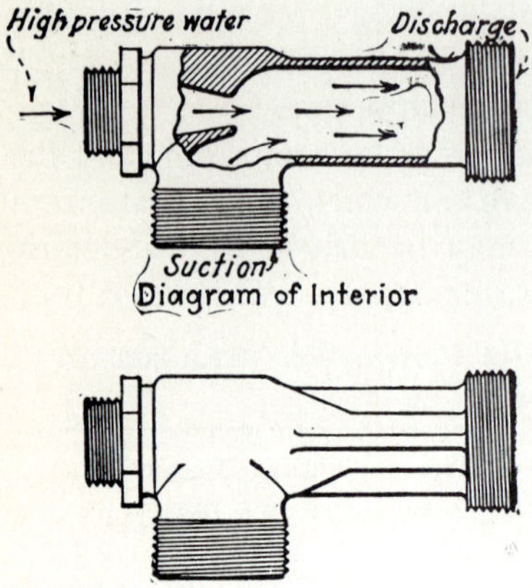

FIG. 11-7. Hydraulic ejector. (*Penberthy Injector Co.*)

TABLE 11-2. Capacities of Penberthy Hydraulic Ejectors

Size number		62	63	64	65	66	67	68
Water-supply inlet, in.		3/8	1/2	3/4	1	1 1/4	1 1/2	2
Suction pipe, in.		3/4	1	1 1/4	1 1/2	2	2 1/2	3
Discharge pipe, in.		3/4	1	1 1/4	1 1/2	2	2 1/2	3
Length, in.		5	6 3/4	8	9	10 1/2	11 1/2	13 1/4
Capacity, gal per hr	40 psi.	500	900	1,500	2,100	3,000	4,200	6,000
	60 to 80 psi.	750	1,000	2,000	3,400	4,500	6,000	9,000

Capacities are given with 10-ft head and include the operating water. The ejector suction should not exceed 22 ft. The total lift, suction plus discharge, can be about 1 ft for every 4 or 5 lb of water pressure.

11-8. Ventilation of Closed Sumps. In the construction of "closed" sumps for the reception of sewage or drainage, ventilation must be provided to avoid the development of back pressures. This is particularly essential when air-displacement pumps are used, since there may be a discharge of sewage or drainage in the house-drainage pipes when the inlet valve to the sump is closed or submerged.

Each sump should be provided with a vent pipe connected to the highest practicable point in the tank unless the tank is of the air-displacement type that requires a completely closed chamber. When the design of the receiving tank is such that the house drain entering the tank may be closed at any time, a vent pipe should be provided on the house drain as near as possible to but below the lowest fixture, soil, or waste-pipe connection. The size and terminal arrangements of the vent pipe should be in accord with the requirements for vents as stated in the Code.

11-9. Rain-water Gutters and Leaders.[1] Rain-water gutters and leaders are provided to catch the rain water falling on roofs or other catchment areas above the ground and to conduct it to some point of discharge, generally the ground surface or an underground drain. It is usually forbidden to discharge such drainage into the sanitary sewer. Materials for rain-water gutters and leaders are discussed in Sec. 11-11.

Rain-water leaders can be placed either inside or outside a building. When inside they are sometimes called *conductors*. When outside they may be called *leaders*. Among conditions favoring an inside location are convenience, safety, appearance, and freedom from freezing. Most rain-water leaders are placed on the outside of buildings, however, because of the lower cost.

Rain water can be collected in a cistern in rural or remote districts where no public water supply is available or soft water is desired. The rain-water leader can empty directly into the cistern, which will be provided with an overflow, or a valve, can be used on the leader to divert any portion of the roof water that is not wanted, such as from the first rush of a storm which is laden with dirt from the roof or when the cistern is already full.

11-10. Sizes of Gutters and Leaders. Sizes of gutters and leaders are listed in Tables 11-3 and 11-4 and of gutters alone are shown in Table 11-5.

All roof or other watershed areas are to be computed as the horizontal projection of the area. Recommendations for gutter, leader, and drain sizes are based on a rainfall of 4 in. per hr unless otherwise stated. If another rate of rainfall is assumed in design, the roof areas are to be divided by 4 and multiplied by the assumed rainfall rate expressed in inches per hour.

The sizes of rain-water leaders, as given in the Hoover Report[2] and shown in Table 11-4, are based on the diameter of circular leaders, and the sizes of gutters shown in Table 11-5 are based on semicircular sheet-

[1] See also L. Blendermann, Storm Drainage System for Flat Roofs, *Air Conditioning, Heating, and Ventilating*, August, 1956, p. 78.

[2] "Recommended Minimum Requirements for Plumbing in Dwellings and Similar Buildings," National Bureau of Standards, July 3, 1923.

metal gutters with the top dimension given. Other shapes should have the same cross-sectional area. Outside leaders above the frost line should be one size larger than inside conductors, as given in Table 11-3. Some codes require that rain-water leaders shall be at least as large as the drains if placed on the steepest slope.

TABLE 11-3. SIZES OF CIRCULAR AND SEMICIRCULAR ROOF GUTTERS AND INTERIOR LEADERS[a]

Maximum area of roof, sq ft		Gutter diameter, in.		Vertical leaders NPC, Sec. 13.6.1[e]	
Hoover Report[b]	"Plumbing Manual"[c]	Hoover Report and "Plumbing Manual"[d]	Of interior leader Hoover Report[b] only, in.	Equivalent diameter, in.[e]	Maximum projected roof area, sq ft
Up to 90	170	3	1½	2	720
91–270	360	4	2	2½	1,300
271–810		4	2½	3	2,200
811–1,800	625	5	3	4	4,600
1,801–3,600	960	6	4	5	8,640
	1,380	7[f]	5	6	13,500
3,601–5,500	1,990	8[g]	6	8	29,000
5,501–9,800	3,600	10			
	6,800	12			

[a] Gutters and leaders of other shapes should have the same cross-sectional areas as are required for the gutters and leaders shown in this table unless otherwise noted. Outside conductors to be one size larger to the frost line.

[b] "Recommended Minimum Requirements for Plumbing in Dwellings and Similar Buildings," National Bureau of Standards, July 3, 1923.

[c] Fall of 1/16 in. per ft or less. "Plumbing Manual," Report BMS66, National Bureau of Standards, 1940.

[d] *Ibid.*

[e] Based on a maximum rainfall rate of 4 in. per hr. If the maximum rainfall rate differs, then the figures for roof area must be adjusted by dividing the figures in this table by the assumed rainfall rate and multiplying by 4.

[f] This size is not shown in the Hoover Report.

[g] This size and larger to be hung with approved wrought-iron hangers.

Recommendations of the Copper and Brass Research Association[1] concerning rain-water gutters and leaders include the following principles:

1. Make leader same size as its outlet, for its full length, to avoid stoppage by leaves, ice, etc.
2. Base runoff computations at 8 in. per hr. [Note that the "National Plumbing Code" and most other standards are based on 4 in. per hr.]
3. Maximum spacing of leaders is 75 ft. A safe rule is 150 sq ft of roof area to 1 sq in of leader area. The figures can be changed to meet local conditions.

[1] "Copper Flashings," 2d ed., The Copper and Brass Research Association, 1925.

4. No leader should be less than 3 in. (circular), where there is possibility of leaves, etc., passing into it; 2-in. leaders are permissible for porches and decks.

5. Size of gutter depends on: (*a*) number and spacing of outlets, (*b*) slope of roof, and (*c*) style of gutter.

6. The best type of gutter has minimum depth equal to half and the maximum depth not exceeding three-quarters of the width. Thus the width becomes the

TABLE 11-4. SIZES OF RAIN-WATER LEADERS*

Type of leader	Area, sq in.	Leader size, in.	Area of roof drained, sq ft	Type of leader	Area, sq in.	Leader size, in.	Area of roof drained, sq ft
Plain round	7.07	3	1,060	Polygon octagonal	6.36	3	955
	12.57	4	1,885		11.30	4	1,695
	19.63	5	2,945		17.65	5	2,650
	28.27	6	4,240		25.40	6	3,810
Corrugated round	5.94	3	890	Plain rectangular	3.94	1¾ × 2¼	590
	11.04	4	1,660		6.00	2 × 3	900
	17.72	5	2,660		8.00	2 × 4	1,200
	25.97	6	3,895		12.00	3 × 4	1,800
					20.00	4 × 5	3,000
Square corrugated	3.80	1¾ × 2¼	570		24.00	4 × 6	3,600
	7.73	2⅜ × 3¼	1,160				
	11.70	2¼ × 4¼	1,755				
	18.75	3¾ × 5	2,820				

* Based on 150 sq ft of roof area for each square inch of leader area.

TABLE 11-5. SIZE OF SEMICIRCULAR RAIN-WATER GUTTERS
("National Plumbing Code," Sec. 13.6.3)

Diameter of gutter, in.*	Maximum projected roof area for gutters of various slopes, sq ft			
	Slope, in. per ft			
	1/16	⅛	¼	½
3	170	240	340	480
4	360	510	720	1,020
5	625	880	1,250	1,770
6	960	1,360	1,920	2,770
7	1,380	1,950	2,760	3,900
8	1,990	2,800	3,980	5,600
10	3,600	5,100	7,200	10,000

* Gutters other than semicircular may be used provided they have an equivalent cross-sectional area.

deciding factor in proportioning its size. There is no reason for a gutter deeper than three-quarters of the width except for ornamental purposes.

Assuming that this proportion is observed the gutter may be referred to by its width only.

A gutter smaller than 4 in. wide is to be avoided. In common practice 4-in. gutters are seldom used for they are difficult to solder and increase the labor cost. The gutter may be the same size as the leader it serves, but, of course, cannot be smaller.

Half-round gutters are most economical in material and ensure a proper proportioning of width and depth.

Safe rules for determining the size of gutters are

1. If spacing of leaders is 50 ft or less, use a gutter the same size as the leader, but not less than 4-in.

2. If spacing of leaders is more than 50 ft, add 1 in. to the leader diameter for every 20 ft (or fraction) additional spacing on peaked roofs.

3. For flat roofs add 1 in. to the leader size for every 30 ft of additional gutter length.

For ordinary residence construction 3- or 4-in. round and 2- by 3-in. or 2- by 4-in. rectangular leaders will generally suffice; 5-in. half-round gutters meet practically every requirement.

A safe system to follow in mill building design is that of the American Bridge Company. Their specifications provide as follows:

Span of roof	*Gutters*	*Leaders*
Up to 50 ft	6 in.	4 in. every 40 ft
50–70 ft	7 in.	5 in. every 40 ft
70–100 ft	8 in.	5 in. every 40 ft

Notes. 1. Round leaders should not be less than 3 in. in diameter.

2. Rectangular leaders should not be smaller than 1¾ by 2¼ in. (This is commonly called "2" sq in.)

3. Gutters should not be less than 4 in. wide.

4. Gutters should have a fall of not less than 1 in. in 16 ft.

5. Scuppers should be provided for all roofs with a parapet wall built around them. This precaution prevents an overloading of the roof due to stoppage of the outlet.

6. All outlets should be provided with screens or strainers.

11-11. Materials for Roof Gutters and Leaders. The "National Plumbing Code" states (Secs. 13.2.1 and 13.2.2):

Inside Conductors. Conductors placed within a building or run in a vent or pipe shaft shall be of cast iron, galvanized steel, galvanized wrought iron, galvanized ferrous alloys, brass, copper, or lead.

Outside Leaders. When outside leaders are of sheet metal and connected with a building storm drain or storm sewer, they shall be connected to a cast-iron drain extending above the finished grade, or the sheet-metal leader shall be protected against injury.

The Code proceeds to state (Sec. 13.5.1):

Material. Roof drains shall be of cast iron, copper, lead, or other acceptable corrosion-resisting material.

Gutters are usually constructed of sheet metal. Cast iron and galvanized wrought iron or steel, brass, copper, and lead are used for leaders within the building or in an inner or interior court or ventilating pipe shaft when these leaders are connected with the house drain or sewer. Where inside conductors are connected to the house drain, sewer material, installation, and other features must meet the limitations applicable to soil and waste pipes. Outside leaders should be constructed of sheet metal with longitudinal corrugations rather than of cast iron or wrought pipe because the sheet metal is not so seriously affected by freezing and by temperature changes. Cast iron is sometimes properly used for a distance

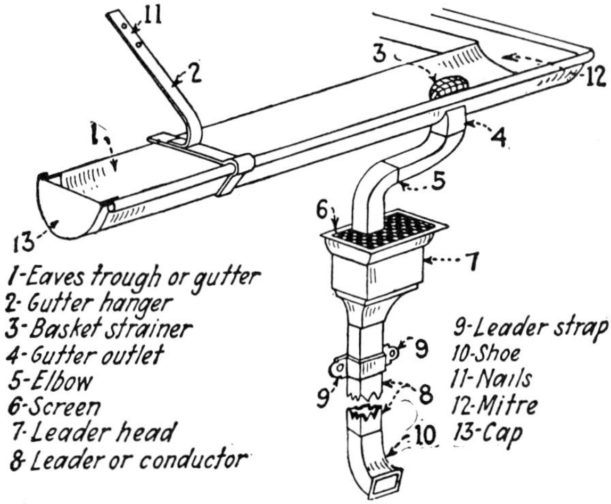

FIG. 11-8. Roof gutter and rain-water leader.

of 1 to 5 ft above the ground surface on account of the hard knocks to which this portion of the leader may be subjected. The entire length of an outside leader should not be made of cast iron, because of damage not only to the pipe itself by freezing but also to the calked joints. The lead will be forced from the joints by the alternate freezing and thawing of the wet oakum beneath it. Wrought iron with threaded joints is better than cast iron with calked joints if a sturdy pipe must be used. Sheet-metal leaders are most commonly used.

The support of the gutter and leader and the connection between the gutter and the leader are important in their erection and maintenance. Installations illustrating good practice and the names used for the parts of sheet-metal gutters and leaders are shown in Fig. 11-8. The thicknesses of sheet metal recommended by different authorities are given in Table 11-6. Recommendations of the Division of Simplified Practice of the U.S. Department of Commerce[1] are:

[1] Simplified Practice Recommendation 29, approved Jan. 1, 1925.

All eaves trough, conductor pipe, shoes, miters, and accessories, including gutters, valleys, ridge rolls, and so on, when made of copper, to be not lighter than 16 oz per sq ft (0.0216 in.).

TABLE 11-6. THICKNESS OF SHEET METAL FOR VARIOUS PURPOSES
(Thickness in inches)

Purpose	Galvanized iron	Copper or brass	Lead
Roof gutters	0.016	0.0215§	
		0.0156¶	
		0.040†	
Outside leaders	0.016	0.040†	
		0.0215§	
		0.0156¶	
		0.064	
Flashing	0.016	0.040†	0.135*
		0.021‡§	0.101‡
Local vents	.....	0.016	
Safes	.....	......	0.101†
Tank lining	.....	0.016	

Authority; * Dibble. † Hoover. ‡ Cosgrove. § Copper and Brass Research Assoc. ¶ Div. of Simplified Practice, U.S. Department of Commerce.

In discussing roof drains, the Bureau of Standards states:

Roof Drains: Body shall be galvanized cast iron with female threaded outlet and have clamping ring for roofing felt; antitilting, removable dome strainer and large sediment chamber; gravel stop; bronze post; brass screws.

Roof Drains for General Use on All Types of Flat Roof: Drains shall be cast iron, japanned, with slag stop, double drainage feature, flashing rings, etc. Domes and flashing rings shall be secured with brass or bronze bolts.

Roof Drains for Use in Cornice Gutters: Drains shall be cast iron, japanned, with outlet for calking.

The connection between the leader and the gutter should be made of brass, copper, or lead pipe and should be corrugated transversely so as to be flexible to allow for inevitable movements between the leader and the gutter. No. 19 B & S gage copper or brass pipe or extra-light lead pipe should be used for the flexible connection. Similar corrugated flexible joints should be installed at intervals along the pipe so that there is at least one fold for each ¼ in. of movement of the pipe. In general, one fold every 50 ft, with never less than two folds, is satisfactory. Outside leaders are subject to wide and rapid changes of temperature, the extreme range being sometimes greater than 125°F.

The opening into the leader from the gutter should flare out at a small angle to twice the diameter of the leader and it should be covered by a strainer to prevent the entrance of leaves, birds, and small animals. The "National Plumbing Code" states (Secs. 13.5.2 and 13.5.3):

> **Strainers.** All roof areas, except those draining to hanging gutters, shall be equipped with roof drains having strainers extending not less than 4 in. above the surface of the roof immediately adjacent to the roof drain. Strainers shall have an available inlet area, above roof level, of not less than 1½ times the area of the conductor or leader to which the drain is connected.
>
> **Flat Decks.** Roof drain strainers for use on sun decks, parking decks, and similar areas, normally serviced and maintained, may be of the flat surface type, level with the deck and shall have an available inlet area not less than 2 times the area of the conductor or leader to which the drain is connected.

Provision should be made to permit water to overflow from the gutter and to prevent its accumulation on the roof in case the leader becomes blocked.

The choice between copper and galvanized- or painted-iron gutters is a matter for the judgment of the designer, who should consider the relative cost and durability of the two materials.

11-12. Installation of Rain-water Leaders. Recommended connections between a roof gutter and a rain-water leader are shown in Fig. 11-8. The right-angle turn immediately below the roof provides flexibility to allow for differential expansion and contraction between the leader and the building. This is important, since the variations of temperature to which leaders may be subjected are sometimes as great as 125°F.

Trapping of rain-water leaders connected to a house drain, house sewer, or other sanitary sewer is essential except where the leader is made of cast iron with calked joints and meets the specifications for stack terminals stated in the Code. Traps on inside conductors should be accessible inside the building. Sheet-metal leaders connected to a sanitary sewer should be trapped because of the possibility of gas leakage through the poorer joints and seams that exist in such pipes. The trap should have a large water capacity and a seal at least 4 in. deep. It may be desirable to discharge the safe waste from a frequently used fixture through the leader to assist in maintaining the trap seal during drought. A sand trap can be used on a leader if no sanitary sewage flows through it. Care should be taken to protect the trap on the leader from freezing by placing it within the building or by burying it below the frost line. Where the trap is buried, access to it should be provided to permit inspection and cleaning.

In order to prevent overflow into a fixture, no rain-water leader should connect to a house drain above any fixture that is less than 5 ft higher than the house drain. The connection between the house drain and the leader should be made by means of not less than one length of cast-iron pipe extending vertically at least 1 ft above grade line. Along driveways without sidewalks the leaders should be placed in niches in the walls, protected by wheel guards, or enter the building through the wall at 45° slope at least 12 ft above grade. The lower end of a rain-water leader is illustrated in Fig. 11-9.

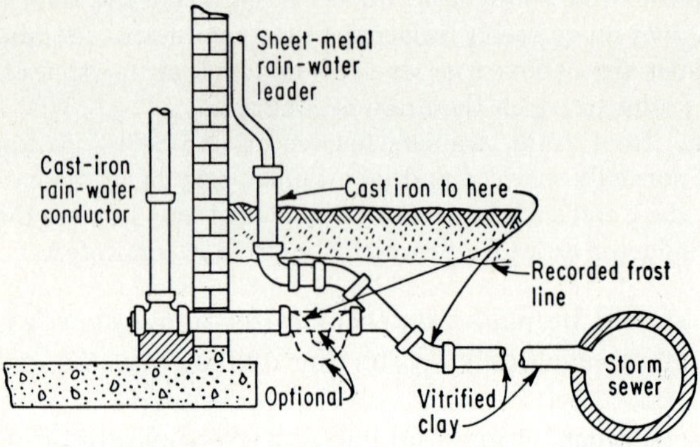

FIG. 11-9. Discharge ends of rain-water leaders and conductors.

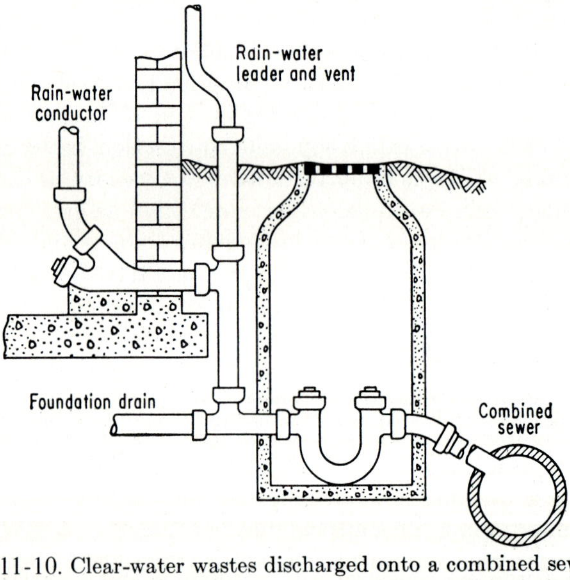

FIG. 11-10. Clear-water wastes discharged onto a combined sewer.

11-13. Clean-water Drainage. Rain-water leaders, yard drains, foundation drains, and other clean-water discharges can be connected to other clean-water drains from the same building to discharge finally into a storm sewer or to a sanitary sewer which is intended to be a combined sewer. It is advantageous to the householder to connect to a combined sewer, since the construction of but one house sewer is thus required, the total cost of plumbing can be reduced, and the heavy discharge of rain water will serve to flush the house sewer. The use of a combined sewer may be prohibited, however, necessitating the connection of rain-water leaders and other clean-water wastes to a storm sewer. Where legally permissible and otherwise feasible, clean-water drains should be trapped and discharged into a combined sewer, as shown in Fig. 11-10. The traps should be protected from frost, as mentioned in the preceding section. As many drains as is feasible should lead into one trap to give greater assurance of maintenance of seal during drought. Trap seals should meet the requirements given in the Code.

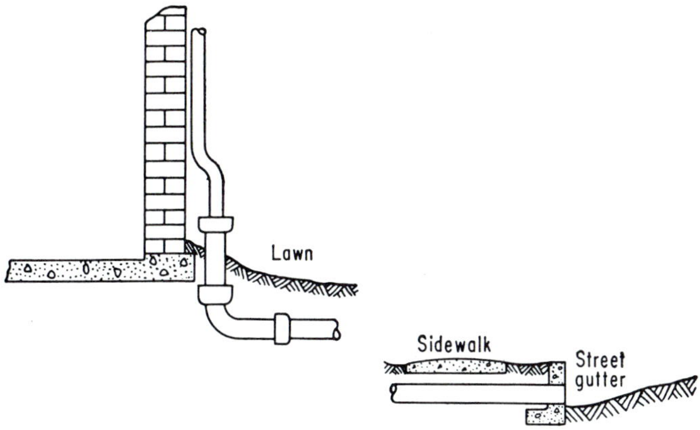

FIG. 11-11. Rain-water discharge into street gutter in a mild climate.

When there is no sewer, rain-water leaders can discharge onto the ground surface provided the slope of the surface will carry the water away from the building or, where freezing does not prevent, the rain-water drain can pass beneath the sidewalk and parking to discharge into the street gutter, as shown in Fig. 11-11.

11-14. Yard and Area Drains. The principal purpose of yard and area drains is to carry off rain water as rapidly as it falls. All roofs and paved areas, yards, courts, and courtyards should be drained into the storm-water sewer system or the combined sewer system, but not into sewers intended for sanitary sewage only. When drains used to carry off storm water are connected to the combined sewer system, they should be effectually trapped unless they terminate as provided for stacks in the Code. A single trap can serve a number of drains, and all traps should be protected from frost by setting within a building or otherwise. Where no sewer is accessible, the discharge can pass, untrapped, under the sidewalk into the street gutter unless prohibited by local ordinance.

Yard and area drains are similar to floor drains except that they should be more substantial in construction and should have a greater capacity for draining off water. The construction must be substantial, and the foundation laid deep, not only to resist the shock of passing vehicles but to resist the heaving action of frost. Types of yard drain are shown in Fig. 11-12. The use of the bell trap shown in the figure is prohibited by the "National Plumbing Code" in Sec. 5.3.6. The traps shown in the figure are made of cast iron, the cover being perforated for the admission of water and the exclusion of debris. If possible, the traps should not discharge into drains carrying sewage, and under such conditions no trap is necessary. Where they must discharge into a sewer, however, they should be trapped with a deep-seal trap with generous water capacity and some clean-water waste should be led into the trap. The trap should be located within a

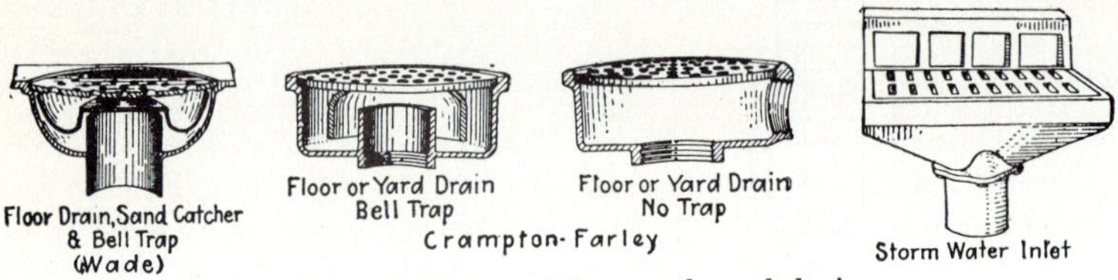

FIG. 11-12. Types of floor and yard drain.

building or beneath the frost line to protect it from frost. It is desirable also to include a sand trap or catch basin.

The capacity of the yard or area drain should depend on the rate of rainfall. An empirical rule suggested in the Hoover Report[1] for estimating the rate of runoff is to consider the rate of rainfall as 4 in. per hr. This will give a rate of flow from each 100 sq ft of impervious area of, closely, 4 gpm. When this is converted into equivalent fixture units, 180 sq ft of drainage area is to be taken as equivalent to one fixture unit. Having determined the rate of flow to be carried by the drain, its size can be found in Table 11-8. Recommendations for drain sizes as made in the "Plumbing Manual"[2] are shown in Table 11-7.

A catch basin should be placed at the inlet to the drain, and the inlet to the catch basin should be protected by a heavy strainer. The capacity of the inlet is as important as the capacity of the drain pipe. If the area of the openings in the inlet strainer is double the cross-sectional area of the inlet drain, no trouble should result from the size of the strainer. Where no catch basin is used or where the catch-basin outlet is not trapped and the drain pipe is connected to the house sewer, drain, or other sanitary sewer, the yard drain should be trapped and access to the trap

[1] "Recommended Minimum Requirements for Plumbing in Dwellings and Similar Buildings," National Bureau of Standards, July 3, 1923.

[2] "Plumbing Manual," Report BMS66, National Bureau of Standards, 1940.

TABLE 11-7. SIZES OF HORIZONTAL STORM DRAINS
(Maximum roof area,* in square feet, that can be drained by horizontal storm drains on various slopes)

Diameter of drain, in.	From "Plumbing Manual," Report BMS66, National Bureau of Standards, 1940				From "National Plumbing Code,"† Table 13.6.2		
	Slope, in. per ft						
	1/16	1/8	1/4	1/2	1/8	1/4	1/2
2	......	......	350	500			
2½‡	......	480	670	960			
3	......	750	1,050	1,500	822	1,160	1,644
3½‡	......	1,100	1,550	2,200			
4	......	1,550	2,150	3,100	1,800	2,650	3,760
5	1,800	2,700	3,600	5,400	3,340	4,720	6,680
6	3,000	4,200	6,000	8,400	5,350	7,550	10,700
8	5,900	8,700	11,900	17,400	11,500	16,300	23,000
10	9,800	15,200	19,600	30,400	20,700	29,200	41,400
12	15,900	24,700	31,800	49,400	33,300	47,000	66,600
15	......	......	......	......	59,500	84,000	119,000

* Based on a rate of rainfall of 4 in. per hr. If the maximum rate of rainfall is more or less than 4 in. per hr, then the figures for roof area, given in this table, must be multiplied by 4 and divided by the maximum rate of rainfall assumed.

† No yard or building drain should be less than 3 in. in diameter.

‡ 2½- and 3½-in. cast-iron soil pipe and fittings and 3½-in. drainage fittings are not generally available.

and the cleaning of the drain pipe should be provided for. The same care should be taken to prevent the freezing of the seal of this trap and for maintaining it as is described for the seal of a rain-water leader in Sec. 11-12. A helpful arrangement in maintaining the seal is to connect the yard drain to the rain-water leader above the trap.

11-15. Cellar Drainage. There may be two types of drain beneath a cellar floor: one to remove ground water that might otherwise press against or seep upward through the floor and one to remove water falling on the surface of the floor. The first may consist of open-joint 4- or 6-in. drain tile laid from a few inches to 18 in. below the cellar floor. Such pipes should be placed not more than about 10 to 15 ft apart on centers, the more porous the soil, the greater the distance apart and the deeper the cover. The pipes should connect and terminate in a down-turned cast-iron quarter bend with its end submerged about 3 in. or less in a water-tight collecting well from which the outflow is a few inches above the face of the inlet elbow, somewhat as shown with a tee in Fig. 11-11. The 4- or 6-in. outflow pipe from this well may join the foundation drain or

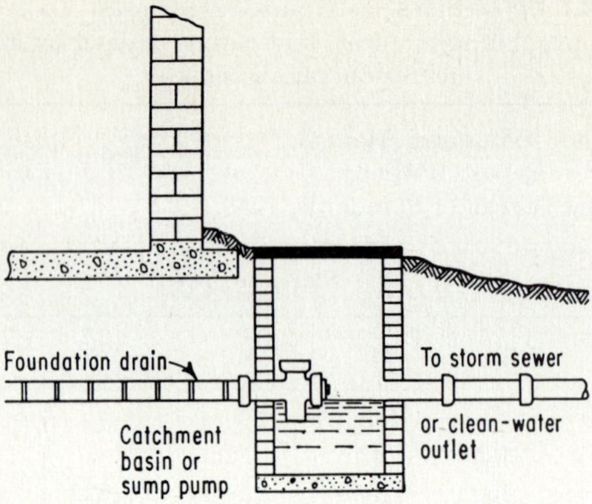

FIG. 11-13. Foundation and subsoil drainage outlet, suitable for a climate where water is below the frost line.

may discharge independently into the storm- or surface-water sewer. It should not discharge into a sanitary or a combined sewer. The collecting well provides a convenient location for a pump, as described in Sec. 11-1, in the event the outlet sewer is too high to receive the flow from the well by gravity.

The surface of the cellar or basement floor should slope at a grade of about ½ per cent, i.e., $\frac{1}{16}$ in. per ft, toward a floor-drain inlet of the type covered in the Code. The distance of flow of water across the floor should be less than about 16 ft. Each floor-drain inlet should be protected by a bell trap or better. The outlet from the trap should discharge into the sanitary sewer. The floor drains may receive any drainage falling on the floor, including the discharge from washing equipment not otherwise connected to the plumbing.

11-16. Subsoil or Foundation Drains. Subsoil or foundation drains are laid close to and outside the foundations of a building to prevent water from entering the building. They are laid somewhat as shown at 58 in Fig. 11-1. The arrangement shown in Fig. 11-1 is objectionable when there is danger of sewage backing through the building sewer. Even a backwater valve may not prevent such difficulties because of clogging of the valve so that it is held open.

The "National Plumbing Code" states (Sec. 13.1.5):

Subsoil Drains. Where subsoil drains are placed under the cellar or basement floor or are used to surround the outer walls of a building they shall be made of open-jointed or horizontally split or perforated clay tile, or perforated bituminized fiber pipe or asbestos cement pipe, not less than 4 in. in diameter. When the building is subject to backwater, the subsoil drain shall be protected by an accessibly located backwater valve. Subsoil drains may discharge into a properly trapped area drain or sump. Such sumps do not require vents.

The foundation drain pipe should be laid at about the elevation of the basement floor and not closer than 3 ft to the foundation wall.[1] If more than 3 ft below the nearest edge of the footing, the pipe should be as far away horizontally as it is vertically below the bottom of the footing. The pipe should be surrounded with gravel or other loose, nonabsorbent material, and the trench should be backfilled with this or similar material to a depth of at least 1 ft and preferably more. The remainder of the trench to within a short distance of the ground surface should be filled with a porous material that will intercept any ground water and carry it to the drain tile. The drain may be laid completely around the building or only on those sides from which ground water is expected. The drain is connected, where possible, to the storm sewer or other surface-water drain. Otherwise it is connected to the building sewer through a trap and a vitrified wye connection where no other point of discharge is possible. If there is a possibility of sewage backing up into the foundation

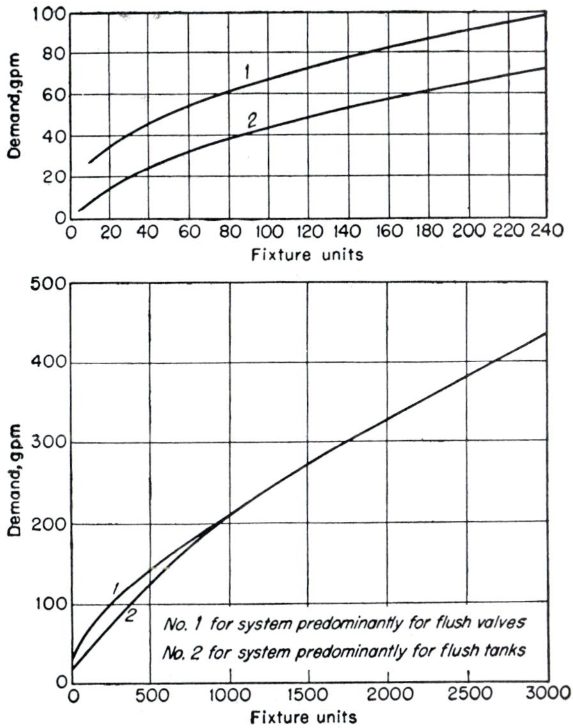

FIG. 11-14. Flow corresponding to various fixture units. (*From "The Uniform Plumbing Code for Housing," February,* 1948.)

[1] "National Plumbing Code," Sec. 2.8.1.

drain, it should be protected by an accessible and otherwise properly located backwater valve. The valve should be higher than the sewage in the building sewer to minimize the possibility of clogging materials in the sewer becoming lodged on the valve seat. The connection of a subsoil drain to a sanitary or a combined sewer should be prohibited.

TABLE 11-8. RATE OF FLOW THROUGH VITRIFIED-CLAY AND CAST-IRON PIPE FLOWING FULL
(Gallons per minute)

Diameter of pipe, in.	Cast-iron pipe							Diameter of pipe, in.	Vitrified-clay pipe						
	Slope								Slope						
	1/16 in. 0.5 per cent	1/8 in. 1.0 per cent	3/16 in. 1.5 per cent	1/4 in. 2.0 per cent	5/16 in. 2.5 per cent	3/8 in. 3.0 per cent	1/2 in. 4.0 per cent		1/16 in. 0.5 per cent	1/8 in. 1.0 per cent	3/16 in. 1.5 per cent	1/4 in. 2.0 per cent	5/16 in. 2.5 per cent	3/8 in. 3.0 per cent	1/2 in. 4.0 per cent
2	8.5	12	15	17	19	21	24	4	60	85	104	120	134	147	170
3	25	35	43	50	55	61	70	6	181	257	315	364	406	445	514
4	52	74	91	105	117	128	148	8	380	540	660	760	850	940	1,080
6	160	225	276	318	356	390	450	10	690	980	1,200	1,390	1,550	1,700	1,960
8	330	470	575	665	740	815	940	12	1,150	1,620	1,980	2,300	2,560	2,810	3,240
10	608	860	1,050	1,220	1,360	1,490	1,720								
12	1,000	1,420	1,740	2,010	2,250	2,460	2,840								

Index